Light Aircraft Maintenance

Light Aircraft Maintenance

A TEXT BOOK FOR AIRFRAME MAINTENANCE

J. E. Heywood
C. Eng., MRAeS, FSLAET

GRANADA
London Toronto Sydney New York

Granada Publishing Limited – Technical Books Division
Frogmore, St Albans, Herts AL2 2NF
and
36 Golden Square, London W1R 4AH
515 Madison Avenue, New York, NY 10022, USA
117 York Street, Sydney, NSW 2000, Australia
60 International Boulevard, Rexdale, Ontario R9W 6J2, Canada
61 Beach Road, Auckland, New Zealand

British Library Cataloguing in Publication Data
Heywood, J. E.
Light aircraft maintenance.
1. Airplanes, Private – Maintenance and repair
I. Title
629.132'5217 TL710

ISBN 0-246-11909-8

First published in Great Britain 1983 by Granada Publishing

Distributed in USA by Sheridan House Inc., 175 Orawaupum St, White Plains, New York, NY 10606, USA

Printed in Great Britain by
Stanley L. Hunt (Printers) Ltd, Rushden, Northamptonshire

Contents

'A book is no less scholarly for being readable,
nor more reliable for being dull.'

C. Northcote Parkinson
('A Law Unto Themselves')

Preface

'The time has come, the Walrus said, to speak of many things.' So wrote Lewis Carroll in his book 'Alice Through the Looking Glass'. The situation does not appear to have changed a great deal with the passage of time; in fact there appear to be more things of which we must speak and write than ever before. Writing a book on aircraft maintenance fifty years ago did not present a great deal of problems since structures and systems were relatively simple and straightforward. Writing this volume today I am conscious of the fact that even the light aircraft is no longer simple and appears to have grown up overnight. Systems that were once the preserve of the airliner are now common on a variety of light aeroplanes and helicopters.

However, any book has its limitations. There will be readers who will ask why I have not touched on this subject or why I have not explored more deeply some other subject. Others will ask why I have not illustrated the system that is fitted to their particular aircraft. The simple answer is that I have tried to provide an introduction to airframe maintenance for a class of aircraft in which variety is the spice of life. To have covered to my own satisfaction all aspects of the subject and all variations of design would have resulted in a volume of such proportions as to give indigestion to the reader and bankruptcy to the purchaser! In addition I have deliberately avoided including material covering basic science and workshop practice since this is readily available in other books such as those produced for the technical colleges. So it is with a knowledge of its inherent limitations that I offer you this volume, a companion to 'Light Aircraft Inspection', as an introduction to a subject in which I have spent my entire working life.

To the young engineer here is a book that will help you to broaden your knowledge and will provide much of the information necessary when studying for aircraft maintenance engineer's licences. To the more senior engineer here is a book that may refresh your memory regarding some of the basics that you should not have forgotten. It is a book that you can place in the hands of the younger men that you supervise in order that they may grow technically and become mature engineers on whom you can rely. To the private owner this book will help you towards a better understanding of your aircraft and will provide you with the basic knowledge to undertake straightforward maintenance tasks on your own aircraft where this is permitted by law.

In no way does this book supersede your aircraft service manual but it does provide a technical background against which the various maintenance procedures may be viewed in perspective. The choice of diagrams has been made with a view to illustrating the text and within the constraints of the material available to me. It does not indicate a preference for one product as against another. Inevitably current engineering practice is linked with real hardware and drawings of hypothetical aircraft rarely serve the purpose in a satisfactory manner. I am indebted to many companies for the use of illustrations appearing in their publications. Whilst I am employed as an airworthiness surveyor with the Civil Aviation Authority it must be understood that the views expressed in this book are entirely my own and should not be taken as representing official policy in any way.

This book would not have been possible without the love and support of my wife Ann. She has not only fed and watered me during long sessions of writing, looked after our four daughters and a busy house, but has also typed the manuscript.

J. E. Heywood
Perth

CHAPTER 1
Basic Workshop Practices

Long before one reaches the stage of wielding a spanner on an aircraft it is self-evident that it must be the right spanner and that there must be the right sort of accommodation within which to wield it. It is obvious that for the vast majority of maintenance some form of covered accommodation must be available. The vagaries of this world's weather are such as to preclude maintenance, other than the more minor tasks, being accomplished outside. Whatever the type or size of hangar it must be dry. A leaky hangar can only lead to water entering an aircraft that has been opened up for maintenance, leading to subsequent structural or functional problems. A further requirement is that of a workshop that is dry, well lit and, dependent upon the climate, heated. Such a workshop will be equipped with metal-topped benches, a cleaning area and a reasonably dust-free floor. A workshop will permit maintenance and repair work to be carried out on various components under satisfactory conditions.

Storage

Anywhere that is used for the maintenance of aircraft has a further need – that of adequate storage. When parts are removed from an aircraft, such as cowlings, inspection panels, seats and components, it must be possible to store them in such a way that they neither get mixed with those from another aircraft nor deteriorate whilst they are stored. A practical solution is found by providing large sets of shelves fitted with polythene curtains.

The provision of individual sets of shelves that can be positioned adjacent to the aircraft being serviced helps to provide the required segregation and also reduces the time taken in storing and retrieving the parts. Psychologically it means that the shelves are more likely to be used if they are close at hand!

However our storage needs do not end there. It is essential that we have suitable storage for serviceable spare parts and components. It is vital that approved and serviceable aircraft parts are completely separated from commercial and unserviceable parts and to achieve this it is usual to have both 'bonded' and 'quarantine' store areas. The size of each may vary from a lockable cupboard the size of a wardrobe to a large complex of buildings dependent upon the size of the maintenance operation. The operation of such a store would be as follows.

On receipt of a delivery of new spare parts that were previously ordered they will be unpacked and checked against the packing note to ensure that all are present and are properly identified. Secondly they will be checked against the original order to verify that they are the correct parts. Thirdly they will be checked against the release documentation to make sure that they have been through the appropriate quality

(971)

Serial No. C 813

1st Copy to Consignee

2nd Copy to File

APPROVED CERTIFICATE

Issued Under Civil Aviation Authority
Reference No. A1/2594/48

Contract No. FT 39281

Sub-Contract Order No.

TO: Perth Aviation Supplies
2 Smith Street,
PERTH, PH1 1DX

AIRWORK LIMITED

PERTH AERODROME - PERTH
SCOTLAND
TELEPHONE No. SCONE 51122/3/4.

Advice and Inspection Note No.

Works Order No. 50/180486

Identification Mark and/or Serial No.	Description – with Part No.	Quantity	Spec. and/or Drawing – with Issue No. or Reference	Cast and/or Batch No.	Details of Test(s) or Test Report Number(s)	Inspector's Stamp
737NJ0418 737NJ0593 737NJ0649 737NJ9587	Light aircraft tyres - 650 x 10 6 ply	6	650C61G1 (1)	W.07792	GTR A23432	
737NJ0661 737NJ0236				W.09795	GTR A23789	
92820547 92841207	Light aircraft tyres - 650 x 10 8 ply	2	650C81-3 (2)	W.05808	GTR A25704	
-	Light aircraft tubes - 600 x 6	6	- (3)	W.07806	MRC X9635A	

Certified that the whole of the materials and/or parts detailed hereon have been manufactured, tested and inspected and, unless otherwise stated above, conform to the full requirements of the appropriate drawings and/or specifications relative thereto in accordance with the conditions of the Air Navigation Order and British Civil Airworthiness Requirements.

Signed ....
For and on behalf of
AIRWORK LIMITED

Date 8th April, 1979.

Fig. 1.1 Certificate of origin.

control processes (fig.1.1). Fourthly they will be physically inspected to ensure that they have not been damaged in transit. Finally, attach a 'SERVICEABLE' label with all the necessary details completed (fig.1.2). Then, and only then, they should be transferred to the bonded store. Only such parts and materials as have been properly checked and found satisfactory may be stored in a bonded store.

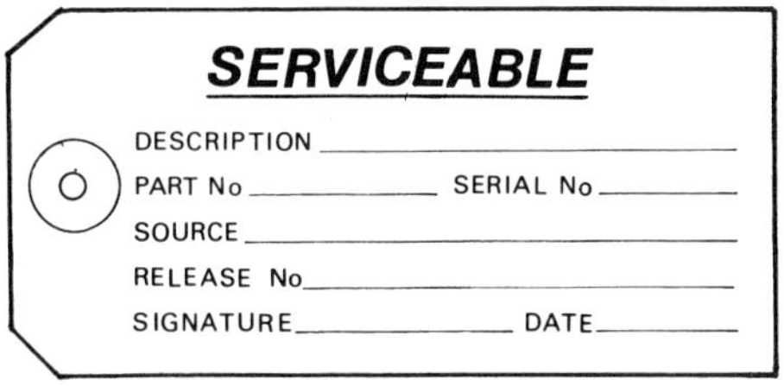

Fig. 1.2 Serviceable label.

If on inspection some of the items had been found to be unserviceable due, for example, to transit damage, then they must not enter the bonded store but should be placed in the quarantine store. The purpose of this store is to provide a store for items that are either unserviceable or are unapproved for one reason or another. A typical example of the sort of item that one would expect to find in a quarantine store would be a hydraulic pump that has been removed for overhaul. This pump would be stored there until such time as it is sent away for overhaul. Of course all items in the store must be satisfactorily identified, usually with a red 'UNSERVICEABLE' label showing that the part is unserviceable (fig.1.3). Items removed for warranty action would also find their resting place in the quarantine store as would items awaiting disposal instructions from their owners.

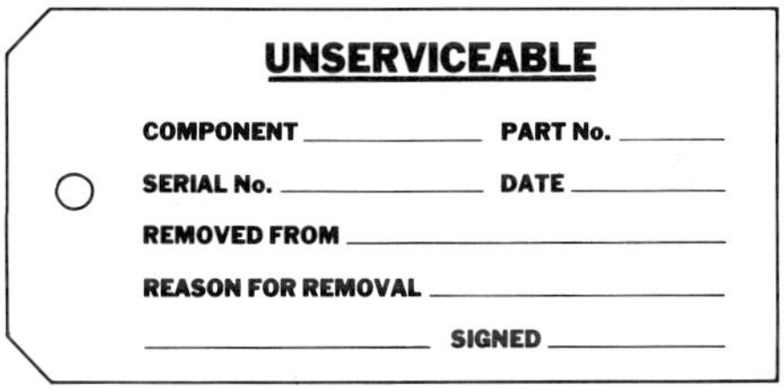

Fig. 1.3 Unserviceable label.

A further type of store is often employed which is generally known as a salvage store. A less euphemistic expression for its contents might be 'useful junk'. It provides a convenient means of keeping both hangar and workshop free of those accumulations of items that might be useful, even if only for repairing ground equipment. Such items as that damaged tailplane, those second-hand wheels and the inevitable hoard of general hardware should be accommodated in this area. Under no circumstances should these be allowed to accumulate around the hangar, in toolboxes or in workshops where they constitute a real hazard to the safe maintenance of aircraft.

Finally there is the genuine scrap and rubbish. Get rid of them outside. Do not allow them hangar room for they are hazardous and take up valuable space that could be better and more profitably employed. Good maintenance begins with good housekeeping. Furthermore a clean and tidy hangar creates a good impression with your customers.

Let us return to our bonded store for a moment. Segregation of stock is not the only feature of a bonded store. It must be possible to relate each part on the shelf to its incoming documentation so that in the event of a subsequent failure the part can be traced right back to the material from which it was made. A simple way to achieve this traceability is to allocate each incoming certificate a serial number and then number each item on the certificate sequentially (see fig.1.1). Thus if the serial number allocated to a given certificate is 971 the reference to be marked on the first item listed on that certificate will be 971/1. The third item on the certificate would be referenced 971/3. The incoming certificates are then filed in serial number order. This type of serial number is often called a batch number and the whole procedure referred to as 'batching'.

A further point that must be taken into consideration in a bonded store is that certain items may have a shelf life. This is a period of storage which must not be exceeded without retest, recalibration or some other procedure to prove the serviceability of the item. In order to ensure that a shelf-life limitation is not exceeded it is essential to have a ready means of ascertaining when the item entered the stores. This can be done by modifying the batch number system slightly, prefixing it with the last two digits of the year and a number representing the month. Thus our batch number would now appear as 7904/971/3, indicating that the third item on the certificate batch-numbered 971 was received in April 1979. This is, of course, only one system among many for the control of parts entering a bonded store but it serves to illustrate the principles involved in operating such a system. In any case old stock should always be used first irrespective of whether or not a shelf life is applicable. This is good stores practice which is facilitated by a batch numbering system such as that described.

It is not only important to control the entry of parts into the store but also their exit from the store. It is important to know which parts were used on a given job for two reasons. One is so that parts fitted to an aircraft may be traced to their source, for airworthiness reasons, and the other is so that the customer may be charged for them.

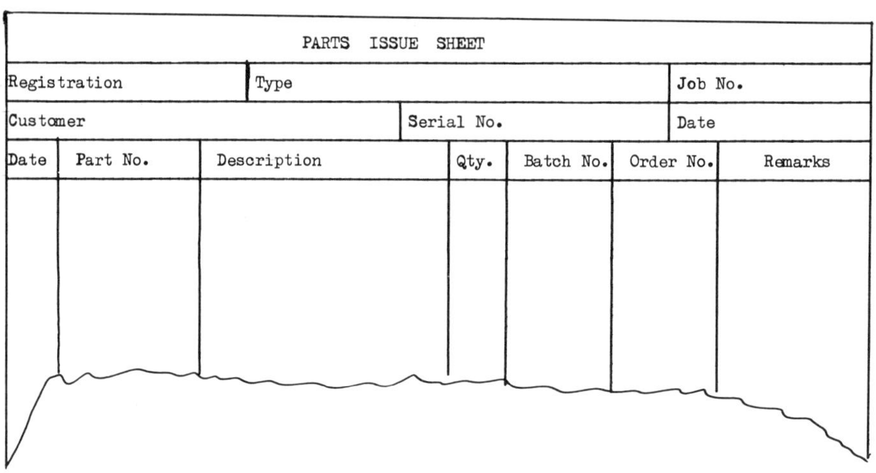
PARTS ISSUE SHEET

Registration	Type	Job No.
Customer	Serial No.	Date

Date	Part No.	Description	Qty.	Batch No.	Order No.	Remarks

Fig. 1.4 Parts issue sheet.

In order to keep account of this aspect of things it is usual to use a parts issue sheet (fig.1.4). As parts are requested, either by written requisition or verbally, so they are entered on the issue sheet. If the parts are supplied from stock the batch number is also entered at the time the parts leave the store. Should the required parts not be in stock they are ordered and the order number is entered in the appropriate column. When the job is completed the spares used are costed and the sheet is filed with the permanent records of the aircraft identified by the job number.

Controlling the movement of spare parts through the store is only one aspect of storage. The conditions under which the parts are stored are very important in maintaining them in airworthy condition. Temperature and humidity must be such that parts will not deteriorate from corrosion. Items which may produce corrosive fumes, such as batteries, should be stored in a separate place altogether. For the same reason the battery shop should be sited as far away from the stores as possible. Flammable materials, that is those with a low flash point, raise special considerations for their storage is usually subject to regulation by various Government agencies. Paints, thinners, fuels, gases, pyrotechnics and solvents all come within this category. Items such as instruments, electronic assemblies and spark plugs are usually stored in a cupboard heated by a 60-watt electric light bulb. Rubber parts, such as hoses, O-rings, and seals should be stored in sealed plastic bags wherever possible. All rubber parts should be protected from direct sunlight and fluorescent light to prevent deterioration from the effects of ultraviolet radiation. Tyres should be stored on suitable racks and turned periodically to prevent distortion of the tread. Components should be stored in their original packings whenever possible. Sheet metal will be useless for aircraft structural work if the surface becomes scratched through careless handling. For this reason it is usual to make wooden racks capable of storing the metal on edge and segregated by thickness and material.

When hoses and electrical connectors have been disconnected, and for such items in store, it is necessary to blank the open ends to prevent the entry of dirt and foreign matter. For this purpose metal or plastic blanks should be used but never paper, rag or masking tape. The type of blank used must make it impossible to fit the mating part without removal of the blank. For example in one known case an internal blank was used on an engine breather fitting and subsequently the breather hose was fitted over the fitting without removing the blank. The resultant build-up of pressure caused failure of the seals in the engine-driven electrical accessories and oil consumption problems.

Workshops

We have already mentioned workshops but let us consider our needs in a little more detail. The floor of a workshop should be painted, or sealed in some other way, in order to keep down the dust that seems inevitable with a concrete floor. Since our workshop will be used for the maintenance of a variety of items it is important to keep it as clean as possible, thus reducing to a minimum the risk of dirt entering systems and components, causing an airworthiness hazard. To assist in achieving this standard the workshop should be partitioned so that one area may be kept for inspection and assembly work and the other for dismantling, cleaning and machining operations. Desirably there should be no direct access from the hangar into the clean area of the workshop and in any case the entrance should be provided with such basic necessities as a doormat. Whilst thinking about the segregation of activities let us say that the workshop is *not* a suitable place for charging batteries. Battery charging must have a room on its own with proper ventilation.

Benches are important since the majority of work will be carried out on them. As a general rule they should be metal-topped with, for example, zinc-coated steel sheet so that they may be easily cleaned and will not be a source of chips and splinters that could find their way into a component being serviced. Several of the benches will need to be equipped with engineers' vices and all should have overhead lighting, preferably of the fluorescent type. Electric power points and compressed air outlets need to be provided to permit the use of power tools. Whilst on the subject of power supplies perhaps it is in order to say a word about the size of compressor that should be chosen. 'Windy' drills and medium spray guns can use anything between 5 and 10 cu.ft. of air per minute at their operating pressure. Too small a compressor is false economy and will only result in unsatisfactory performance and high maintenance costs.

Cleaning parts

Cleaning facilities deserve a word at this point. Whatever fluids are used for cleaning the tanks should be kept covered when they are not in use. This will reduce the loss of solvent and prevent contamination by airborne dirt and dust. There are a number of proprietary solvents on the market which offer a much reduced fire risk by comparison with kerosene. They also have the added advantage of being free of water in suspension, the presence of which can cause corrosion of some parts. Petrol (gasoline) is a very good solvent but the fire risks associated with it are so high as to make its use inadvisable. Those fluids sold as dry cleaning solvents are usually suitable for general degreasing and cleaning. Where it is required to remove paint or carbon deposits it will be necessary to use a chlorinated hydrocarbon type of solvent. In this case evaporation of the solvent from the tank is usually prevented by having a sealing layer of fluid on top of the active chemical. The depth of this layer must be periodically checked by dipping with a glass tube, placing the finger over the open end and then withdrawing it from the tank. The depth of the seal will be clearly visible in the tube, as shown by the change in colour, and should then be adjusted as required.

The most convenient way of using mineral-based degreasing solvents is to store them in a drum which is surmounted by a basin in which the cleaning is done. The solvent is fed through a flexible nozzle by an electric pump and drains back to the drum by gravity. Such a system allows the dirt to settle at the bottom of the drum. When cleaning a component the dirt is worked away using the solvent and a one-inch paint brush, or a stiff bristle brush where the deposits are harder. Ball and roller bearings should only be cleaned using a small amount of fresh solvent in a dish, such as a domestic baking tin. Even a small amount of dirt from solvent that has already been used is sufficient to ruin an otherwise serviceable bearing. Likewise any brush used to clean bearings must be absolutely spotless.

One point that may need consideration where any cleaning operations are carried out is that of ventilation. Many solvents, when used for prolonged periods, can be a health hazard and an extractor fan should be installed unless the natural ventilation is sufficient.

So much for our basic workshop. We shall meet the various tools and pieces of equipment that are used as we examine each of the various aspects of the technology that we call maintenance.

Ground handling

With light aircraft the majority of ground movement is conducted by the simple expedient of manpower. But even in the application of so basic a means some care is necessary if accidents to personnel and damage to aircraft are to be avoided.

Before moving any aircraft always check that the ignition switches are in the *off* position. Too many accidents have happened through neglect of this simple precaution. As a follow-up to this the throttle should be closed and the mixture control should be in *idle cut off*. Do not take a chance with any aircraft – it just is not worth it.

Basically two things are involved in moving light aircraft – pushing and steering. Firstly all pushing should be on the strong points of the aircraft such as the mainplane leading edge, the landing gear and the hub of the propeller. Do not push on tailplanes, control surfaces or the tips of the propeller blades. Secondly steering should be accomplished by using the correct steering bar for the aircraft. Like most ground equipment steering bars have a hard life and become worn. Make sure that it locks properly or you may find yourself with a runaway aircraft. A centre line painted on the ground into the hangar is a great asset when positioning an aircraft, especially when you are on your own. But do ensure that you have adequate clearance for the wing tips. It must sound like a statement of the obvious to say such a thing but unfortunately many light aircraft have experienced 'hangar rash' simply because someone did not check. Helicopters can often be parked in tighter spaces than fixed-wing aircraft but all the more care must be taken in untangling the inter-meshing rotor blades. Additionally it is often harder to judge the diameter of a rotor than it is to judge the span of a mainplane.

The functions of pushing and steering can be carried out mechanically by using one of the small tractors designed for the purpose. These are excellent units but, like any machine, will only do as they are told. So the observations made above with regard to manhandling apply equally well when using a tractor. One additional point is that the clutch must be engaged smoothly in order to prevent unnecessary strain on the aircraft.

Picketing

Most light aircraft are provided with strong points by means of which the aircraft may be tied down. If an aircraft is to be left outside overnight then it should really be tied down, particularly in the case of the smaller models. A pair of chocks just will not hold the aircraft if the wind gets under it. For example, consider an aircraft with a stalling speed around 45 knots; the wind does not have to be a lot faster than that for the aircraft to leave the ground, especially as it will be empty.

How should we tie down our aircraft? First of all it should be parked into the direction of the prevailing wind unless a weather forecast indicates otherwise. Then the parking brake should be applied firmly and the wheels double chocked. There should be a chock in front of and behind each main wheel and they should be tied together round the wheel. The aircraft can then be tied down to adequate stakes driven into the ground at an angle. Each tie down anchor should provide a minimum holding power of approximately 3 000 pounds and this can be achieved with stakes driven into dry, grass-covered ground. However, soakage of the ground will reduce this strength drastically and under these conditions two stakes, crossed at 90°, should be used. Ropes made from synthetic materials, such as nylon or polypropylene, are preferable since these do not shrink when wet (fig.1.5). Ropes should be tied so that they are just snug but not tight, except that natural ropes

Strengths of Common Tiedown Ropes

Size (diameter) inch	Minimum Tensile Strengths for Twisted Ropes (pounds)			
	Manila	Nylon	Dacron	Polypropylene
1/4	600	1 500	1 440	1 300
5/16	1 000	2 400	2 200	1 900
3/8	1 350	3 400	3 120	2 750
1/2	2 650	6 200	5 500	4 200

Braided ropes have a minimum tensile strength that is approximately 30–40% less than twisted ropes.

Fig. 1.5 Tiedown details.

should have about an inch of slack to allow for shrinkage. Obviously the security of the aircraft is no better than the knots that are used and these should be of the non-slip variety such as a round turn with two half hitches or a bowline. Typical angles for the ropes and their locations are shown in the diagram (fig.1.6). Finally make sure that the controls are locked, preferably at the control surfaces, so that they will not be damaged by being blown on to their stops. The tendency of the wings to develop lift in very high winds can be negated by securing wood battens (50 x 50 mm) on top of

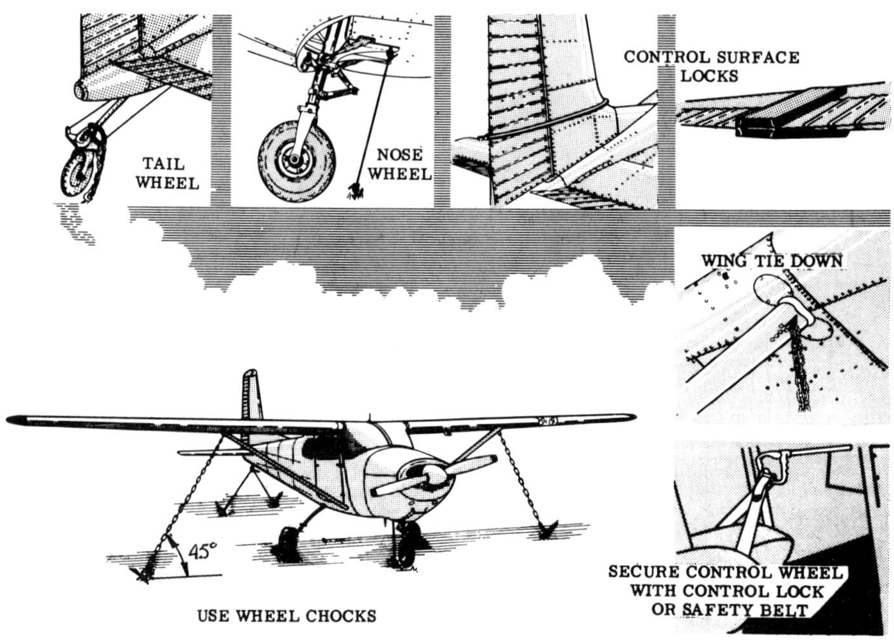

Fig. 1.6 Tiedown ropes.

the leading edges to act as a spoiler. Sandbags may also be used but care must be taken to ensure that the wing structure is not overloaded. Their best use is to secure the wooden battens. Do not forget to fit covers to all openings such as pitot, static and fuel vents. Always use the strong points that are provided by the manufacturer since the use of other parts of the aircraft could result in distortion of the airframe under some conditions.

Jacking and trestling

On many occasions during the maintenance process it is necessary to lift the aircraft clear of the ground in order to permit, for example, the landing gear to be retracted or removed. The exact method of achieving this varies with the individual aircraft and various methods will be illustrated, but some points are common to all jacking operations and will be enumerated first.

Irrespective of the type of jacks to be used ensure that they have sufficient load capacity and total lift for the aircraft concerned. The safety locks should be checked in order to see that they operate correctly. The general serviceability of the jacks should be checked to ensure that all the parts are present, correctly attached and capable of functioning properly. It may also be necessary to check the level of the hydraulic fluid. At the same time as the jacks are checked do not forget to check the tail trestle, or weight, when it is necessary for such an item to be used. The correct jack pads should be fitted to the aircraft and the aircraft configuration checked to ensure that any limitations imposed by the manufacturer are not going to be exceeded. For example, it may be prohibited to jack the aircraft when it is above a certain weight or fuel load. If subsequent work is going to require the defuelling of the aircraft it should be done outside the hangar before jacking takes place. Then consideration should also be given to other aircraft and a location chosen in the hangar where the jacked aircraft will not prove an obstruction. Aircraft should not be jacked outside or where they are likely to be caught by the wind.

Having paid due regard to all the foregoing considerations the jacks may be placed under the jack pads and extended until they are properly engaged. At this point walk round the aircraft and check that jacking will not cause the aircraft to foul any other aircraft or ground equipment. Make a final check that the jacks will not foul the landing-gear doors, either while jacking or during a subsequent retraction test. With one person positioned on each jack lifting can now proceed with the jacking taking place as evenly as possible and a watchful eye being kept for any signs of unexpected nose or tail heaviness. As the jacking proceeds screw down collar-type safety locks so that they are always within two threads of the top of the jack body. Where pin-type locks are used always insert the pin in the lowest hole as soon as possible. Raise the aircraft high enough to permit subsequent retraction tests but no higher, secure the safety locks and remove the jack handles (fig.1.7). Where a tail trestle is used, make sure that there is no likelihood of the aircraft leaving it during subsequent maintenance operations that would reduce the weight at that end of the aircraft. Up to this point no one should be underneath or in the aircraft.

Lowering the aircraft is substantially the reverse of the above procedure. Take the same safety precautions to ensure stability of the aircraft and clearance with any obstructions. Additionally, ensure that the landing gear is fully extended and locked by checking each one physically in addition to the normal indications. Raise each jack slightly to release the safety locks and then lower slowly and evenly. Do make sure that the aircraft weight is taken fully by the landing gear before trying to remove the jacks. Failure to do so could result in an expensive hole in the structure.

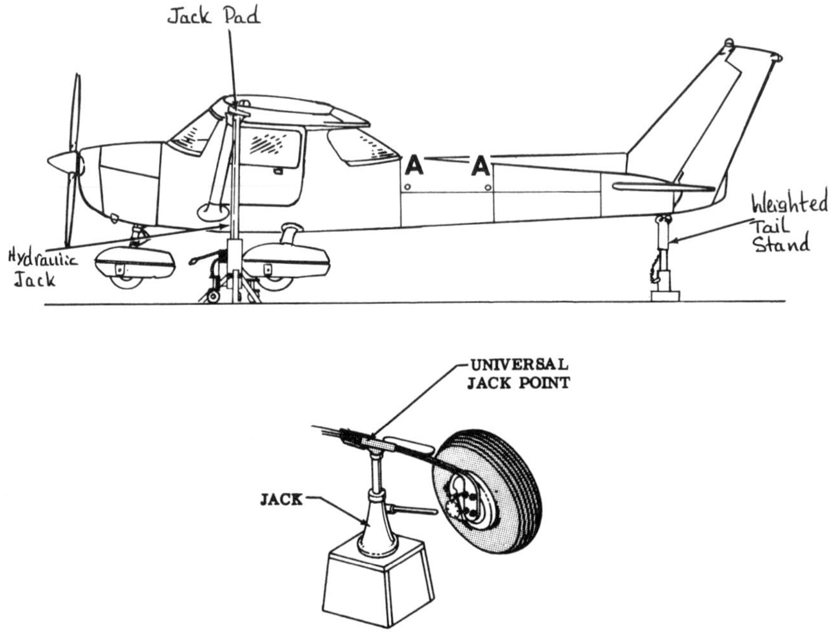

Fig. 1.7 Jacking.

Jacking an aircraft to change a single wheel follows the same principles as already enumerated. Before starting to jack double chock the other wheels and, where possible, apply the parking brake. But never rely on a parking brake in the absence of chocks. If a bottle jack is being used and it is too short in the closed position then a *substantial* block of wood may be used as a packing piece (fig. 1.7). However it must be wider than the jack, the height must not exceed the width and it must be in one piece. Bricks or building blocks must not be used. To try to jack under other conditions is to risk damage to the aircraft or the people involved.

When jacking an aircraft with spring steel landing gear it will be necessary to do this in several stages, raising and lowering as necessary to compensate for the sideways movement of the aircraft. Where a jack does not have a safety lock, such as a bottle jack, then it is a sensible precaution to put a large block of wood under the axle after the wheel has been removed just in case! With one wheel jacked no one should enter or stand on the aircraft.

So much for some of the basic procedures associated with the handling of aircraft and their parts. We ought now to take a look at one or two basic engineering practices that are in constant use in the world of aircraft engineering.

Because of the nature of the operating environment aircraft designers go to great lengths to stop parts unscrewing. The two most common devices in use are lock wire and the split pin and for that reason we must be thoroughly familiar with their correct use.

Wire locking

Wire locking, or safety wiring, is a most positive and satisfactory method of safetying screwed parts. Whilst there are standard methods of wire locking in common use it may be that a manufacturer will specify a particular method of locking a particular item. If this is so there will be a good reason for it and it must be done that particular way. Lock wire is available in different materials and again where one is specified it must be used. For general use stainless steel wires are preferred to galvanised soft iron wires since they can be used in all locations without fear of deterioration over extended periods. Copper wire should only be used when actually specified and then only in the size specified. The usual application for thin copper wire is sealing of valves or switches that are only required for emergency use and then it is essential that it can be broken when necessary.

When choosing a safety wire for general use make sure that it is a reasonable fit in the hole in the part. Using too thin a steel wire in a soft component will result in it pulling right through the edge. Two popular lockwire sizes are 0.032 in. dia. and 0.040 in. dia. in stainless steel to the American specification MS 20995. The smaller wire is suitable for all normal airframe requirements whereas the larger wire should be used to meet heavy duty needs, for example, internal lockings on engines and larger flying control turnbuckles. The diagrams illustrate some common types of wire locking. Before attempting to wire lock a threaded part it must be tightened to its proper torque value. If it is necessary to align a castellated nut with a hole in the bolt then it should be tightened to the low side of the torque range. It may then be further tightened, but staying within the upper torque limit, in order to align the holes. If it is impossible to achieve the necessary alignment the whole operation should then be repeated with another nut.

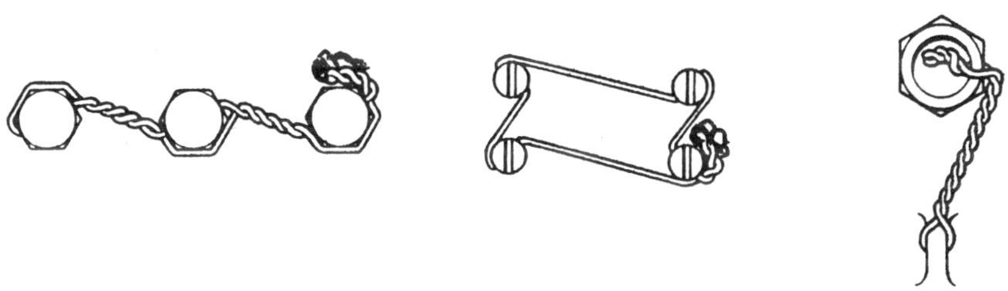

Fig. 1.8 Wire locking.

The direction of the wire should always be such as tends to tighten the part; the twists should be tight and even with the wire between the nuts as taut as possible without risk of breakage (fig.1.8). The twisting may be done by using one of the special tools available but many still prefer to use a pair of duckbill pliers. The lay of the wire should be such that it is tangential to the bolt head as shown. The final end should be about ½ in. long (about six twists) and then bent back on itself so that it does not present a hazard to personnel nor catch in cleaning rags. If the wire is overtwisted then it will suffer from fatigue which will weaken it and prevent it from achieving its purpose. For the same reason lock wire must never be reused. The continuous method of wire locking should, generally speaking, be used only for small screws and even then the length of wire should not exceed 24 ins. Satisfactory locking of pipe couplings is sometimes difficult to achieve and care must be taken to ensure that the lay of the wire is not closer than 45° to the axis of the union.

To remove lock wire it should be cut at each loop and then gently pulled free. Care must be taken to pull the wire in line with each hole or light alloy components can be damaged by steel wire. Having removed the wire do not just throw it on the hangar floor since it is quite capable of puncturing the tyres on a light aircraft. Likewise do not leave it lying around since it has a habit of finding its way into places where it can cause an airworthiness hazard. Lock wire is probably best cut with a pair of diagonal cutters since these will reach into places with relatively limited access. If stainless steel wire is being used then it will be necessary to use a pair of cutters that have hardened jaws or damage to the cutting edges will result. However this type does tend to be rather brittle and can chip at the end if dropped on to a concrete floor from a work stand.

Split pins

These are also called cotter pins and, as with lock wire, are available in different materials. Both stainless steel and cadmium-plated carbon steel are in use, the latter being the most popular for general use. However, only the stainless steel pins are suitable for high temperature applications and are distinctly preferable on marine aircraft. Split pins are supplied in varying diameters measured in fractions of an inch but it should be noted that British and American practice is different with regard to split pin diameters for different thread sizes. It is important that the pin is a snug fit in the hole or it will rattle with vibration and, in a severe case, wear right through. A further point to notice is that there are some applications that require a 5/64 in. dia. pin and a 1/16 in. dia. pin is not satisfactory.

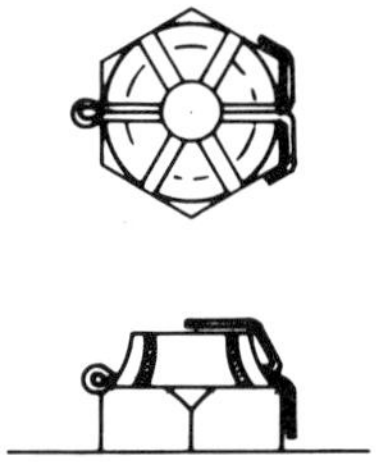

Fig. 1.9 Split pins.

Split pins may be cut to length on installation and the legs folded round the nut in one of the two ways illustrated (fig.1.9). However it is important that the legs are not folded too sharply or there is a danger of them cracking and subsequently failing. Where the split pin has been installed on a moving part checks should be made to ensure that sufficient clearance exists with other parts as the movement takes place. Where a large nut is used on a hollow axle an alternative to using one long pin is to use two short pins fitted diametrically opposite. These are fitted from the outside and the legs spread inside the axle. Nuts must be tightened to their proper torque before fitting split pins and the remarks made in connection with lock wire apply equally well in this context.

Split pins may be removed by straightening the legs and then pulling them out with a pair of pliers. They should never be reused for the same reason that safety wire should not be reused. Discarded split pins, and cut ends, should always be disposed of so that they cannot present a hazard.

Thread inserts

Thread inserts made from diamond-section hardened steel wire are to be found in many parts made from light alloys, especially when it is necessary to remove the mating fastener at regular intervals. However the inserts are also widely used for repair purposes when the female thread in a component has stripped. In many cases it is possible to make the repair without extensive dismantling, thus saving a considerable amount of time and expense.

The selection of the proper tools and inserts must be made from the manufacturer's data sheets and the installation made in accordance with the following procedure. The damaged hole is drilled out to the correct size and then tapped using a special tap appropriate to the size of the insert. During both these operations consideration should be given to greasing the drill or the tap in order to prevent the swarf entering other parts of the assembly. A thread gauge is available for checking the size of the prepared hole prior to installing the insert. The insert is installed into the hole by the use of a tool which slightly compresses, or pre-winds, the insert to facilitate the job (fig. 1.10). The insert is placed in the chamber with the tang towards the nozzle and the mandrel pushed forward through the insert to

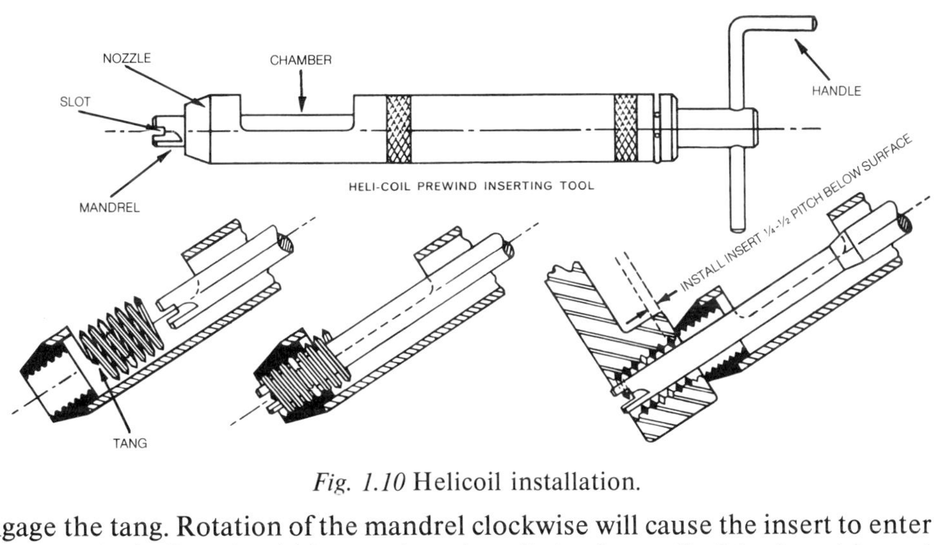

Fig. 1.10 Helicoil installation.

engage the tang. Rotation of the mandrel clockwise will cause the insert to enter the nozzle and winding should be continued until the first coil of the insert is ready to emerge at the other end. The tool should now be placed squarely over the tapped hole in the component and the insert wound into place. The last coil of the insert should be ¼ to ½ pitch below the surface. The tang may now be broken off using the special tool. The punch is inserted into the installed insert until it touches the tang. A smart blow with a hammer will break off the tang without disturbing the insert.

On some occasions access may be restricted and the standard pre-wind tool cannot be used. In these cases a simple mandrel can be used although rather more care is necessary when starting the insert into the tapped hole. The diameter of the mandrel must be such that it is a clearance fit in the minor diameter of the installed insert and may be made to fit any convenient tool, such as ¼ in. square or hexagon drive. An installed insert may be removed by tapping the extractor tool into the insert and rotating it anticlockwise.

Torque loading

One of the first things learnt by an apprentice fitter of some years ago was the answer to 'How tight is tight?' By the time he had stripped a thread or two he was developing a feel for the subject. In these days we are more scientific and the torque spanner has largely taken over from the educated feel of the fitter. In many cases stress levels are much higher and therefore there is less latitude in the apparently simple job of tightening a nut or bolt. The resistance of a fastener to fatigue or the resistance of mating parts to fretting can all depend upon correct installation and tightening. A commonly used method of predetermining the stress in a fastener is by measuring the torque required to tighten the nut under a specific set of circumstances. This will include whether or not the thread is lubricated or plated, the thread form, the material of the fastener and so on. Thus it is important to read the small print associated with the torque tables published in the manufacturers' manuals to ensure that the torque is applied under the right conditions (fig.1.11).

RECOMMENDED NUT TORQUES

THE TORQUE VALUES STATED ARE POUND-INCHES, RELATED ONLY TO STEEL NUTS ON OIL-FREE CADMIUM PLATED THREADS.

FINE THREAD SERIES

TAP SIZE	TENSION TORQUE		SHEAR TORQUE	
	STD (NOTE 1)	ALT (NOTE 2)	STD (NOTE 3)	ALT (NOTE 2)
8-36	12-15		7-9	
10-32	20-25	20-28	12-15	12-19
1/4-28	50-70	50-75	30-40	30-48
5/16-24	100-140	100-150	60-85	60-106
3/8-24	160-190	160-260	95-110	95-170
7/16-20	450-500	450-560	270-300	270-390
1/2-20	480-690	480-730	290-410	290-500
9/16-18	800-1000	800-1070	480-600	480-750
5/8-18	1100-1300	1100-1600	660-780	660-1060
3/4-16	2300-2500	2300-3350	1300-1500	1300-2200
7/8-14	2500-3000	2500-4650	1500-1800	1500-2900
1-14	3700-5500	3700-6650	2200-3300	2200-4400
1-1/8-12	5000-7000	5000-10000	3000-4200	3000-6300
1-1/4-12	9000-11000	9000-16700	5400-6600	5400-10000

NOTES

1. Covers AN310, AN315, AN345, AN363, MS20365, MS21042, MS21044, MS21045 and MS21046.
2. When using AN310 or AN320 castellated nuts where alignment between the bolt and cotter pin slots is not reached using normal torque values, use alternate torque values or replace the nut.
3. Covers AN316, AN320, MS20364 and MS21245.

Fig. 1.11 Typical torque value table.

The value of torque is given by multiplying the force applied by the radius at which it is acting and so it will be a compound unit. In the imperial system the most common units are pounds-feet (lb/ft) and pounds-inches (lb/in). To convert pounds-feet to pounds-inches it is only necessary to multiply by 12. In the metric system the kilogramme-metre (kg/m), or a multiple of it, has long been the unit. However with the advent of the ISO system this has been replaced with the metre/decanewton (m/daN).

There are two main types of torque spanner in common use, the preset type and the direct reading type. In the former the spanner can be preset to the value required and indicates when this is reached by 'breaking', releasing or operating some form of indicator. A direct reading torque spanner indicates on a dial or scale the torque being applied at any moment. Obviously a torque spanner, like any precision tool, is only as good as its calibration and this should be periodically checked and a record kept. This may be carried out on one of the special rigs manufactured for the purpose but there is a simpler method for those not so fortunate. We may apply our knowledge of moments to assist us in checking the calibration. First clamp the square drive in the protected jaws of a vice. Known weights may be hung from the handle and the torque calculated by multiplying the applied force by the distance between the centre of the square drive and the point at which the weights are attached. (Due allowance must be made for the torque produced by the weight of the torque spanner.) This should be done at several points within the range of the torque spanner, compared with the reading on the spanner, and a note made of any deviations.

Having assured ourselves of the correct conditions and the calibration of our torque spanner the torque should be applied slowly and smoothly until the required value is reached. The torque spanner should be slowly released and the process repeated to ensure the reading is correct. Where a castellated nut has to be lined up with a hole in the bolt it is usual to torque to the bottom end of the torque range and then tighten until the next slot is aligned. However in critical situations this is not permissible so check your maintenance manual. If a fastener is taken beyond the maximum torque limit it should be removed and scrapped. Where self-locking nuts are used it may be necessary to determine the torque caused by the resistance of the locking element and to add this to the desired torque value to obtain the proper reading on the torque spanner.

Aircraft being what they are it is not always possible to use a standard socket and offset or extended sockets must be used (fig. 1.12). Where an extension is at 90° to the

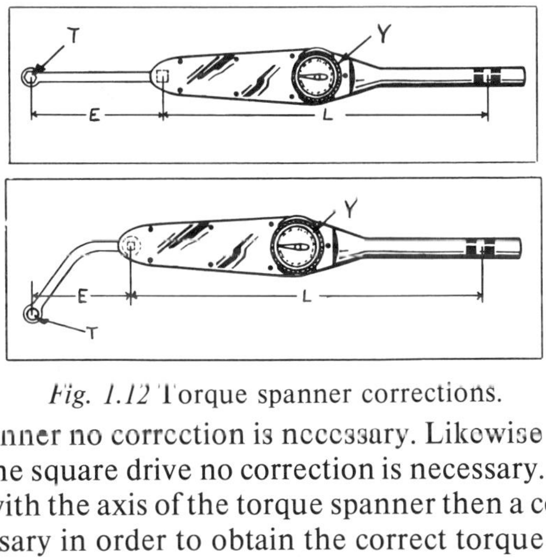

Fig. 1.12 Torque spanner corrections.

axis of the torque spanner no correction is necessary. Likewise if the extension is in line with the axis of the square drive no correction is necessary. However should the extension be in line with the axis of the torque spanner then a correction to the scale reading will be necessary in order to obtain the correct torque at the fastener. This may be summarised in the formula:-

$$Y = \frac{T \times L}{L + E}$$

where: Y= the reading required on the torque spanner scale
T= the desired torque at the fastener
L= the effective length of the torque spanner
E= the effective length of the extension.

This holds good for the case where the extension lengthens the torque spanner and it will be seen that the scale reading will be less than the desired torque.

Should the extension shorten the torque spanner the formula changes as follows:

$$Y = \frac{T \times L}{L - E}$$

Thus the reading on the scale of the torque spanner will be higher than the desired torque.

Crack detection

From time to time it will be necessary to examine parts for signs of cracks. Obviously this can be done visually, aided if necessary by a x10 magnifying glass. However an extremely sensitive and very popular method of crack detection is that using penetrant dye. Used properly this process is quite capable of demonstrating the existence of cracks whose total volume is only of the order of one millionth part of a cubic centimetre. If you cannot visualise that size it would be a crack 0.001 mm wide by 1 cm long and 0.1 mm deep (approximately 0.00004 in x 0.4 in x 0.004 in). The process can be used to find flaws in almost any material, including aluminium, steel, brass, some plastics and ceramics, provided that the defect is open to the surface. The penetrant is drawn into the defect by capillary action and subsequently is drawn from it and shown up on the contrasting white developer.

The process is simple although great care must be taken if the results are to be reliable. There are five basic steps in the process.

(1) **Cleaning** - It is absolutely essential that the part is spotlessly clean before going any further. All paint, scale and hard carbon must be removed with a suitable stripping agent. Following this the penetrant remover/cleaner may be used to remove any traces of oil or surface dirt, spraying it on to the surface and wiping it off with a clean cloth. This should be repeated until no contamination is visible on the cloth.

(2) **Applying penetrant** - Spray the surface of the part with the penetrant, making sure that it is evenly wetted. If the penetrant runs into droplets the part must be cleaned again to remove the contamination. The penetrant should be allowed to remain on the surface for a period of 10 to 20 minutes. An ambient temperature above 15°C is necessary for the best results but it should not exceed 32°C.

(3) **Removal of excess penetrant** - Wipe the surface with a clean dry cloth or tissue. Follow this with a further cloth or tissue, which should first be moistened with penetrant remover. Finally dry with yet another clean cloth. It is very important to ensure that the surface is thoroughly cleaned or spurious indications will result.

(4) **Developing** - Shake the aerosol can vigorously until the agitators are heard to rattle inside it. Then spray a thin, even film on the surface. If the film is too thick small indications will be masked.
(5) **Inspection** - Defects are revealed as bright red marks on a white background. Cracks show as lines or as a series of red dots if they are fine. Some estimation of the depth of the defect can be obtained by the speed with which the penetrant spreads through the film of developer. At least 10 minutes should be allowed before inspection for very small defects.

This is a reliable process provided care is taken. It is paramount that the part is clean before starting and that the removal of the penetrant is completed thoroughly. Both these operations are particularly difficult on such items as threaded parts and rough castings.

Corrosion

This is not the place for a full treatment of the subject of corrosion, particularly in view of the various excellent articles and books available on the subject. However a basic working knowledge of what happens, how to identify it and how to rectify the problem is essential to any aircraft engineer. Due to the nature of the environment in which we operate the problem will not go away but we must use every means at our disposal in order to control it. This means constant vigilance and attention to the smallest indications of corrosion if serious problems are to be avoided at a later stage.

First of all, let us take a quick look at the causes of corrosion and the forms it can take. Corrosion can be caused by direct chemical attack or electrochemical attack but in either case the metal is converted to one of its compounds, e.g. oxide, hydroxide or sulphate. In an electrochemical attack the difference in electrical potential between two dissimilar metals, or even two areas of the same metal, in the presence of moisture forms an electrical cell in which the anode is attacked. In direct chemical attack the metal is in contact with a corrosive fluid, e.g. salt water, battery acid, caustic cleaning solutions, urine and toilet fluids and the like. The corrosion always starts on the surface or at an edge but the subsequent progress may be external on the surface of the metal or internal through the thickness of the metal. In aluminium and magnesium alloys the corrosion has a whitish-grey appearance whilst ferrous metals show the characteristic reddish-brown colour.

Where surface corrosion is present the surface of the metal will be roughened, or perhaps pitted with a powdery deposit. Painted surfaces will often reveal the presence of corrosion of the underlying metal by a blistering of the paint film caused by the products of corrosion. Corrosion within a riveted joint will often be detected by a slight swelling of the joint between adjacent rivets. Corrosion caused by the contact of dissimilar metals can be difficult to find since it is usually out of sight in the joint area, which may be a bolt hole in a spar. Another form of corrosion results in a breakdown of the grain boundaries in the metal and can progress through a component with very little evidence visible on the surface. This intergranular corrosion can cause the metal to break up into flakes when the condition becomes known as exfoliation. One cause of this corrosion is a lack of uniformity in the material stemming from the original heat treatment. Any component showing signs of intergranular corrosion must be scrapped since it is impossible to determine the full extent of the damage.

Another form of corrosion related to intergranular corrosion is known as stress corrosion. It is found where there are sustained stresses in a material coupled with a corrosive environment. Typical examples of parts subject to cracking as a result of stress corrosion are torque links with pressed-in bushes or taper-thread grease nipples, overtightened tubing nuts and items with shrink or press fitted bushes and bearings. Fretting corrosion occurs when two heavily loaded parts are able to move slightly and is a form of abrasion between the mating surfaces.

A basic knowledge of the various causes of corrosion is essential to prevention. From what we have already seen we can deduce that if we can keep fluids away from the part then, in most cases, we are well on the way to controlling the problem. So the most important single factor in controlling corrosion is cleanliness. It is essential to remove accumulations of dirt and other deposits as soon as possible since these will hold moisture in contact with the structure of the aircraft. Regular and proper lubrication of moving parts will also exclude moisture. Regular and careful inspection will bring to light any corrosion whilst it is in its early stages and before it becomes a serious problem. Corrosion-prone areas that should receive special attention include:

exhaust stain areas
bilges
flap shrouds
toilet compartments
battery stowages
wheel wells
inside of pressure skins

Any inspection should ensure that drain holes in the structure are clear and that drain valves on pressurised aircraft are functioning properly. Where dissimilar metals are involved they should always be assembled using a zinc chromate based sealing compound in order to exclude moisture from the joint. Wing structures and firewalls should always be properly sealed in order to preclude the entry of exhaust gases into the structure. This is particularly important on twin-engined aircraft.

So much for prevention. But we live in a real world and from time to time the need arises to rectify a part in which corrosion has already become established. The first step in rectification is to establish the extent of the corrosion and to do this satisfactorily it will be necessary to paint strip the surface. The paint surrounding the affected area should be masked off whilst paint stripper is applied. The paint stripper must be compatible with the material to which it is being applied and also with any resins or sealants employed. Some paint strippers will weaken joints in a bonded structure or dissolve the sealants used in the skin joints of pressure cabins and therefore only a stripper known to be satisfactory should be used. Do not forget to take all the necessary personal safety precautions when using these materials including gloves, apron and goggles. When the area is clean and the extent of the corrosion certain we may proceed with its removal.

In the case of ferrous materials mechanical removal with a wire brush, steel wool or wet and dry paper is usually satisfactory. However in the case of highly stressed parts, such as landing gear axles and similar parts great care must be taken and only fine abrasives must be used. Failure to heed this warning can result in scratches from which a crack may subsequently start. Proprietary phosphoric acid solutions may be used following mechanical cleaning but should never be used on highly stressed parts where they can cause embrittlement. In the case of aluminium and magnesium alloys as much corrosion as possible should be removed with a stiff bristle brush followed, if necessary, by mild abrasive compounds or fine sandpaper used wet. A chromic/sulphuric acid based solution, used in conjunction with a stiff bristle brush, is a suitable treatment for removing heavier corrosion deposits. Great care should be taken to ensure that such a solution does not enter closed structures or riveted joints and that it is thoroughly washed off after a few minutes' use. If this is not done properly the last state may be worse than the first.

With all the corrosion deposits removed it is necessary to make a thorough inspection in order to assess the damage that has occurred. A few scattered shallow pits need not be cause for concern but should the pits form a line or be deeper it will be necessary to measure their depth. The depth of each defect should be compared to the thickness of the skin and the percentage reduction of thickness calculated. For example, where the material is 0.025 in. thick and the corrosion pitting is measured as 0.005 in. there is a $0.005 \times 100/0.025 = 20\%$ reduction in thickness. The actual reduction in strength may be out of all proportion to the reduction in thickness and will be further complicated if this is a pressurised skin. The repair schemes published by the manufacturer should be consulted in order to find the appropriate action.

When the necessary repairs have been completed the surface must be protected in order to prevent a further occurrence of the corrosion. This is the subject of Chapter 5.

A basic tool kit

The choice of tools is always a personal thing with every engineer having his own preferences. However cheap tools are almost always a poor investment and only lead to frustration. This is particularly true of Phillips type screwdrivers, for the cheap varieties simply fail with a tight screw.

CHAPTER 2
Airframe Rigging

Locations

When an aircraft is built at the factory the assembly drawings are used as a basis for building assembly jigs that will ensure the correct location of each part in the airframe structure. When an aircraft has been involved in an incident or accident, or following a major repair, it is necessary to ensure that the primary attachments are in the correct places. In order to measure these locations accurately it is first necessary to be able to define them in all three planes. This is achieved by drawing lines parallel to the three axes of the aircraft and measuring these from an arbitrary datum for each axis (figs2.1 and 2.2). Thus any specific point in the aircraft can be defined as the junction between three lines, one in each plane. The three lines are known as fuselage stations, buttock (butt) lines and water lines and are parallel to the normal, the longitudinal and the lateral axes respectively. Where buttock lines are on a wing they are often known as wing stations. In order to indicate which side of the datum a particular point lies a convention is adopted as follows:

fuselage stations forward of the datum are water lines below the datum are	negative
fuselage stations aft of the datum are water lines above the datum are	positive

Buttock lines (wing stations) are always positive but it is usually indicated whether they are port or starboard if an assembled aircraft is being discussed.

The datum may be a point on the airframe, such as the forward face of the engine firewall but in many cases the datum is in space ahead of the fuselage. The sole reason for this arrangement is to make all the arms positive for the purpose of weight and balance calculations. In many helicopters the datum is the centre line of the main rotor mast. The actual station number is the distance in inches (or millimetres) from the datum, so once the datum is located any other point on the airframe may be checked for its correct location.

Levelling

Before measurements can be carried out on an airframe it is usually necessary to level the aircraft. This aligns the aircraft so that the longitudinal and lateral axes are horizontal. The aircraft is jacked up, as previously described, until the wheels are clear of the ground. The jacks are then adjusted as necessary to bring the aircraft to the level position. The level position may be determined in a number of ways but two of these are more common than others. With many small single-engine aircraft a level is laid across the seat rails to check the lateral level and then on two screws, or

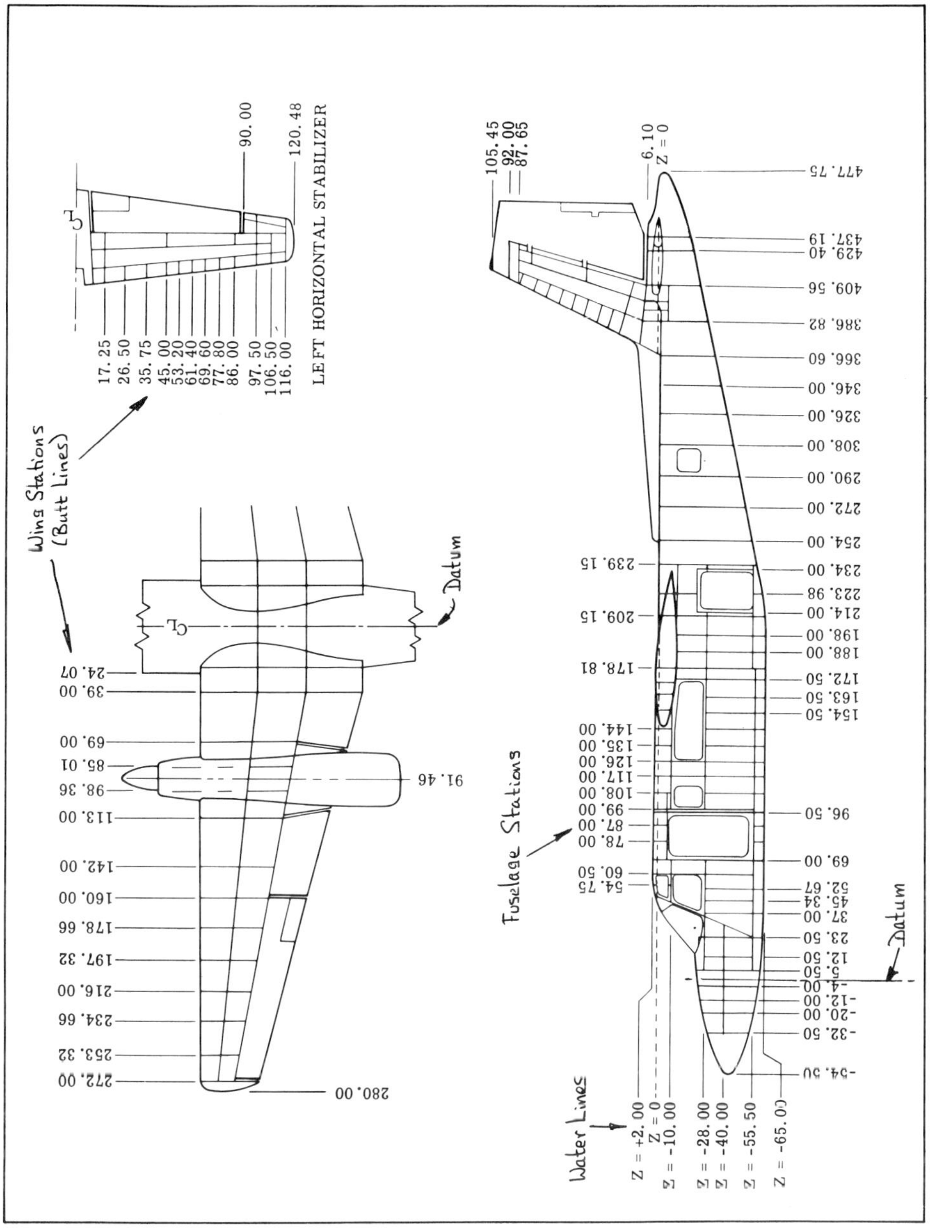

Fig. 2.1 Station diagrams – aeroplane.

pegs on the side of the fuselage to check the longitudinal level. In the case of steel tube fuselages it is common practice to find small pegs welded to the frame in suitable locations. In some cases it will be necessary to lay a straightedge across the levelling points and then place the level on top of it. Once the aircraft is level the jacks

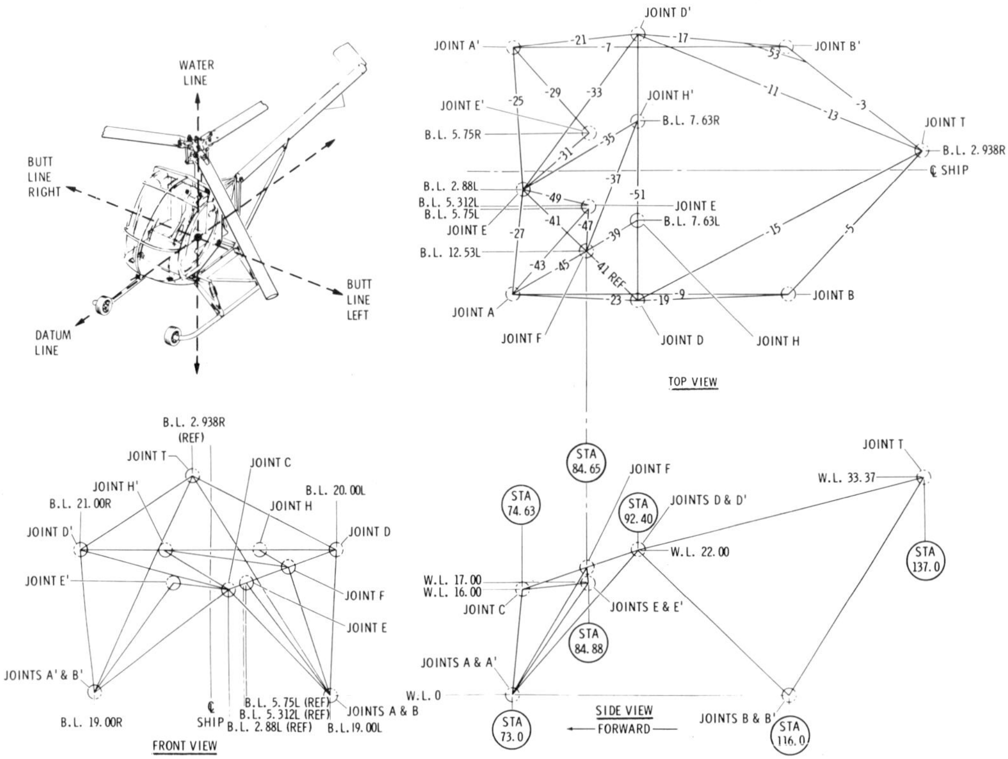

Fig. 2.2 Station diagrams – helicopter.

should be locked and steady trestles placed in position at the wingtips but clear of the actual wing.

With some tricycle landing gear aircraft it will also be necessary to have a similar trestle at the tail. This level position is also known as the rigging position. This is the basic position for the aircraft for carrying out all dimensional checks.

The method of carrying out specific checks varies from one aircraft to the next as dictated by the construction of the aircraft and the preference of the manufacturer. However the principles are the same.

Equipment

Although some aircraft may need special tooling there are certain basic tools with which the engineer must be familiar. Once the aircraft has been levelled angles are measured using a precision inclinometer. In this the bubble is accurately centred and the angle read off using a vernier type scale. Distances are measured using a steel tape. The points between which measurements are to be taken may be projected on to the floor by dropping a plumbline from the point on the aircraft. Sometimes measurements are taken directly and then it is usual to use a spring balance in line with the tape in order to ensure that the same tension is used for each measurement.

Dihedral

The dihedral angle of the main planes is commonly checked by laying a straightedge along the main spar and placing an inclinometer on top. However in the case of many tapered wings the dihedral angle will be that of the centre line of the spar and to measure this it will be necessary to use a tapered wedge instead of a straightedge. In many cases it is only necessary to know that both wings are the same and then comparative measurements may be made from the upper surface of the main spar.

Incidence

Two parameters must be considered in the context of incidence, namely the angle between the chord line of the root rib and the longitudinal axis and also the wing twist. The latter, also known as wash-out when the incidence decreases towards the tip, may exist over only part of the wing and therefore several measurements may be necessary.

In order to measure incidence some manufacturers give dimensions to enable the chord line to be established. This is drawn on the rib, a straightedge clamped in place and the angle measured with an inclinometer. In many cases it will be found that the chord line coincides with two jig holes that are used when the wing is being built. In such cases the straightedge may be placed across two bolts fitted in the holes. The

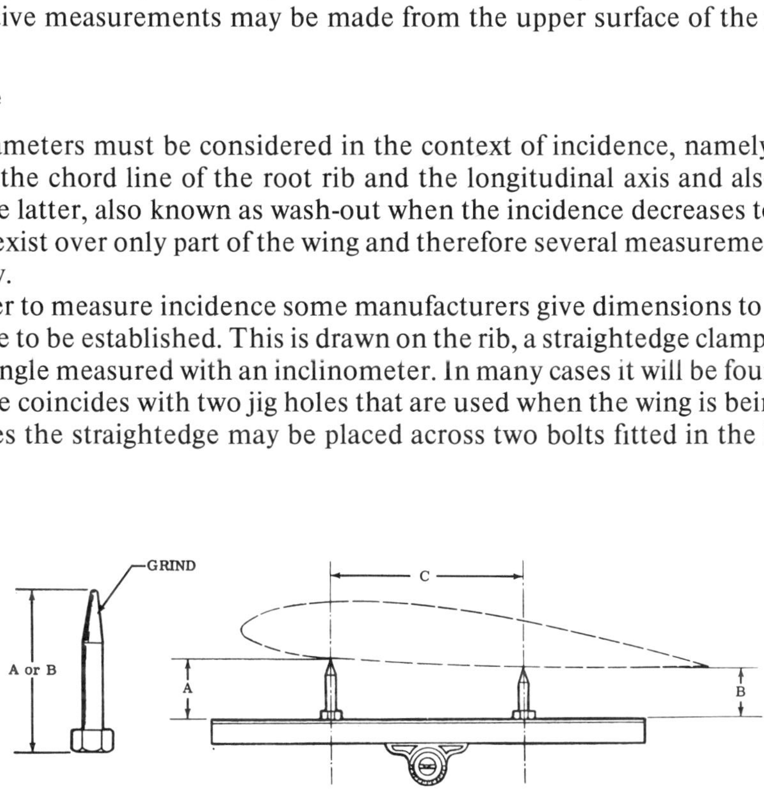

A	B	C	WING STATION
2.09	1.00	28.75	39.00
1.55	1.00	24.50	207.00

Fig. 2.3 Typical wing incidence measurement.

process should be repeated at the wing tip and the difference between the two measurements will be the total twist in the wing.

Where the rib chord line is not accessible or where it is necessary to measure the incidence in the middle of a wing it will be necessary to make a special jig (fig.2.3). This consists of a rigid bar equipped with pegs of specific lengths for a particular wing station. Such dimensions should be given in the repair section of the service manual. In many cases the peg lengths are arranged so that the bar remains level if the correct twist is present in the wing. The same principles of measurement may be applied to the tailplane.

With some aircraft it is extremely difficult to establish the chord lines or datum planes when the aircraft is assembled, as would be the case where a heavy landing is involved. In such cases a long piece of clear PVC tubing nearly filled with water may be used to transfer a point from one part of the structure to another in a different location. This behaves as a U-tube and the water always finds its own level. Measurements should always be made from the bottom of the meniscus.

Engine thrust line

After a major repair it may be necessary to establish that the engine thrust line is in the correct location (fig.2.4). Often this line is offset by a degree or two to compensate for torque or airflow reaction. With the engine removed the difficulty lies in measuring to a point in space. This problem can be solved by clamping plywood boards across the engine mount points. The relationship between the mount points and the crankshaft will be found on the engine installation drawings and one manufacturer, at least, publishes these in the service manual. Using these data draw the thrust line on each board with a pencil. It now remains to check this line against the appropriate datum and this is one occasion when a water-filled PVC tube is most useful.

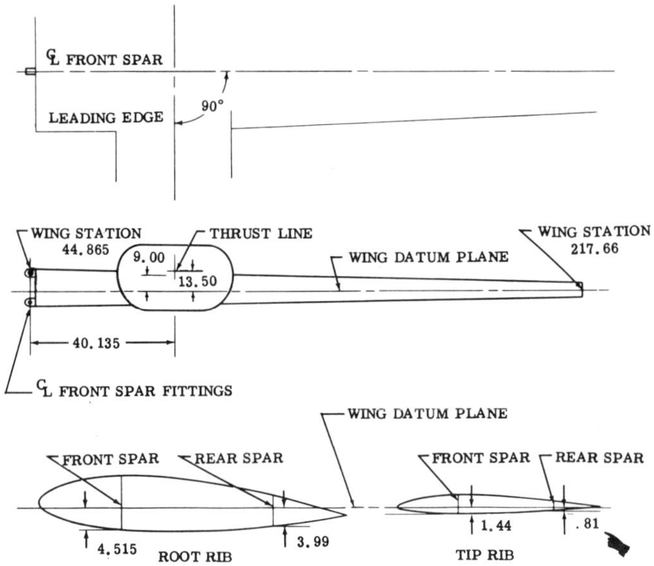

Fig. 2.4 Engine thrust line measurement.

Symmetry

An assembled aircraft must be the same both sides of the longitudinal axis or it will not fly straight and level with 'hands off'. To verify that this is so, or to determine if any damage has taken place in a heavy landing, it is necessary to carry out a symmetry check. In making a symmetry check measurements between certain defined key points are made and compared with those made on the other side of the aircraft. They must agree within a tolerance specified by the manufacturer (fig.2.5).

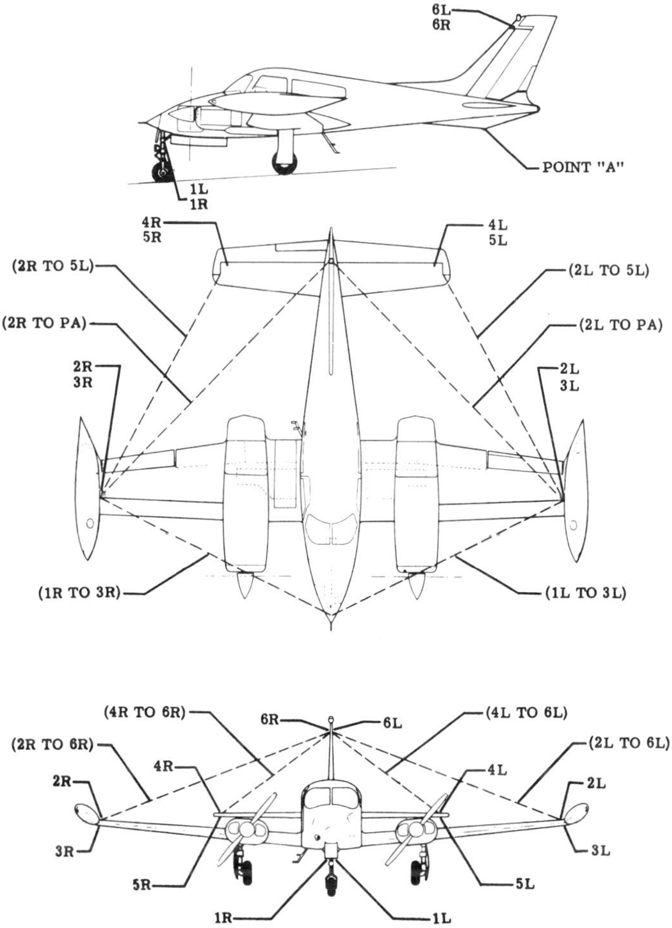

Fig. 2.5 Symmetry measurement.

Usually these measurements are made in two planes. But it is also necessary to ensure that the fuselage is not bowed in any direction. To check for lateral bow plumb bobs are dropped from central points on a selection of main frames and their locations marked on the floor. A chalked line is then used to establish whether or not these all lie in the same plane. To establish a central point where none is obvious a plumb bob may be dropped from each side of the fuselage, just touching the skin, and the locations marked on the floor. The two points are then joined by a line. The mid-point of the line will be the central point of the frame above it. The same procedure may be followed utilising primary attachment points such as lift strut, wing and undercarriage fittings.

Alignment of a fuselage in the vertical plane is a little harder. As with all dimensional checks the fuselage must be in the rigging position. The height above the floor of a known point, such as a wing or landing gear attachment, may now be measured. The relationship between this point and any other points in the vertical plane can be found by reference to the water lines on the station diagram or assembly drawings. Measurements can now be made to establish the accuracy of fuselage alignment.

Although this chapter has largely been written in the context of a fixed-wing aircraft the principles apply equally well to a helicopter structure. It is just as important for all the key points to be in correct alignment on a helicopter as it is on a fixed-wing aircraft.

Nice as it is to know that your aircraft is the shape the manufacturer intended there is not a great deal that can be easily adjusted on the modern aircraft. If discrepancies exist after, for example, a heavy landing, then major repairs are likely to be involved. However, one manufacturer does make it possible to make small adjustments to the mainplane incidence in order to correct small flying faults. The rear wing attachment is made using two eccentric bushes, thus permitting the trailing edge of the wing to be raised or lowered slightly. It is important that both front and rear bushings are rotated simultaneously, otherwise a shear load will be imposed on the bolt. A word of caution. If both left and right bushings are at their mid-point, the thick side must be either outboard or inboard for *both* sets of bushings. If one set faces inboard and the other outboard one wing will be moved forward and the other aft at the wing tip. The result of this will be a significant reduction in the rate of climb. The condition would, of course, be revealed by a symmetry check.

Older design, fabric covered, high-wing aircraft often offer more scope for adjustment. In this type of design there are two lift struts for each wing, one being attached to the front spar and one to the rear spar. Both lift struts are adjustable in length by means of a screwed fork end. When rigging this type of assembly the front lift strut is adjusted first to obtain the correct dihedral and then the rear one is adjusted until the washout is correct.

The ultimate in the possibility of adjustment is the strut and wire braced biplane. Not only may all four wings be individually rigged but also the centre section between the upper pair of wings. The rigging of the centre section must be accomplished, first making sure that it is in line with the centre line of the fuselage, level laterally and correct for incidence and stagger. Then the lower wings may be rigged followed by the upper wings. It can be a formidable process on some aircraft.

The first step in rigging a centre section is to locate it so that the longitudinal centre line of the section is directly above the centre line of the aircraft. To do this adjust the crosswires (a) and (b) (fig.2.6) until they measure the same length from pin to pin. This can be checked by using a set of trammels and measuring from the centre of each pin. If the length of wire (a) is greater than that of wire (b), loosen wire (b) and tighten wire (a). When adjusting wires that are in tension always loosen one wire before tightening the opposing one. When wires (a) and (b) are adjusted to equal length the centre section is in the middle of the aircraft, provided that the struts (c) and (d) are not adjustable. If these struts are adjustable they should be adjusted until they measure the same from bolt to bolt before attempting to adjust the crosswires.

To check the accuracy of the work drop plumblines from the upper wing fittings and measure the horizontal distance from the lines to some structural part of the fuselage such as the lower wing fittings. Obviously, if the aircraft is level laterally and the centre section is in the middle the distances will be the same on each side.

Next, place a straightedge directly over the front spar in the centre section and place a level or an inclinometer on it. If the centre section is not level the struts (c)

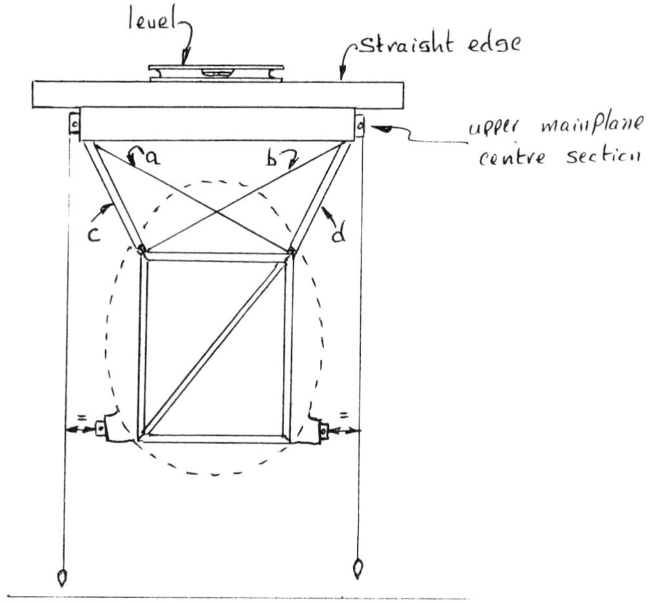

Fig. 2.6 Biplane rigging (1)

and (d) should be readjusted. If they are of equal length and the centre section is not level the misalignment may be due to improper stagger. To check this drop a

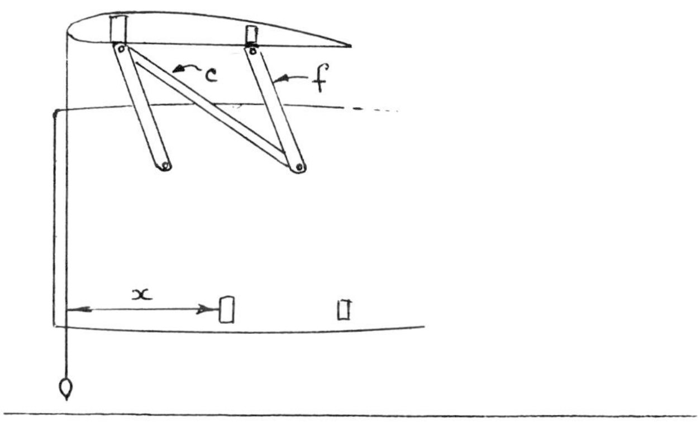

Fig. 2.7 Biplane rigging (2).

plumbline over the leading edge of the centre section near each end (fig.2.7). The distance from the plumbline to the wing fitting (x) should be the same on both sides of the fuselage. If the stagger is greater on one side than on the other the stagger struts (e) should be adjusted. For example, if the stagger is too great on the left-hand side, the strut (e) should be shortened; if too little, the strut should be lengthened.

It should be noted that any adjustment of the stagger on one side will affect the horizontal level of the centre section and therefore it should be rechecked after making any adjustments.

When the stagger, the level of the front spar of the centre section and the location about the centre line are correct the final operation is checking the incidence, which must be the same on each side of the centre section. This may be checked by any of the means which we have previously described. Adjustments of the strut (f) on either side will bring the angle of incidence to the correct value. Further, having made this adjustment, all the other adjustments must be rechecked to ensure that they are still correct. It cannot be overemphasised how important is the correct rigging and alignment of the centre section, for if this alignment is not correct it will be impossible to rig the wings correctly and the flight characteristics of the aircraft will be impaired.

With the centre section correctly rigged we can move to rigging the lower wings. The lower wings are rigged for correct dihedral by adjusting the length of the landing wires, making sure that the dihedral is the same for both wings. The dihedral of the upper wings is adjusted by lengthening or shortening the front interplane strut.

The next step in rigging a biplane is to adjust the incidence of both lower wings by lengthening or shortening the rear landing wire. With the lower wing correctly rigged a check should now be made of the stagger between the upper and lower wings by dropping a plumb bob from the leading edge of the upper wing. It will then be possible to measure the distance between the plumb-bob line and the leading edge of the lower wing. Should this prove to be incorrect it may be adjusted by lengthening or shortening the diagonal interplane strut. Again both wings must have identical measurements. Finally check and adjust the incidence of the upper wing by shortening or lengthening the rear interplane strut. After the adjustments have been made go round and recheck each one of them before finally tightening any bolts or lock nuts. When it is certain that each of the measurements is correct the flying wires can be tightened to the proper tension. After these final adjustments and checks have been made all strut adjustments should be locked, and the tie rods streamlined and safetied.

To summarise then, the proper sequence for rigging the wings for a biplane may be outlined as follows:

1 Set the dihedral angle of the front spars of both lower wings.
2 Rig equal incidence angles in both lower wings.
3 Set the stagger.
4 Adjust or check the dihedral angle of the front spars of the upper wings.
5 Rig the incidence of the upper wings.
6 Set any wash-in or wash-out that may be necessary.
7 Recheck all the adjustments and safety all tie rods, struts, and adjustments.

Whatever the aircraft and whatever the adjustment always double check that it is right. Recheck the measurements after all the bolts and lock nuts have been tightened and check any other measurements that could be affected by the adjustment that has been made. Make a table or diagram and record all the final measurements on it, sign it and date it. This will provide a permanent record of the aircraft rigging on that date and will provide a basis for comparison at any subsequent time.

Needless to say the proof of the adjustment is in the flying. Whenever adjustments are carried out the aircraft must be test flown to verify that all is as it should be.

CHAPTER 3

Metal Airframe Structures

Basic construction

Over the years many different materials have been used for the construction of airframes. However we shall only consider airframes which are constructed from metal, in one form or another, since this is by far the most common form of construction today. Metal construction of fuselages tends to fall into two basic divisions, semi-monocoque (fig.3.1) and steel tube (fig.3.2).

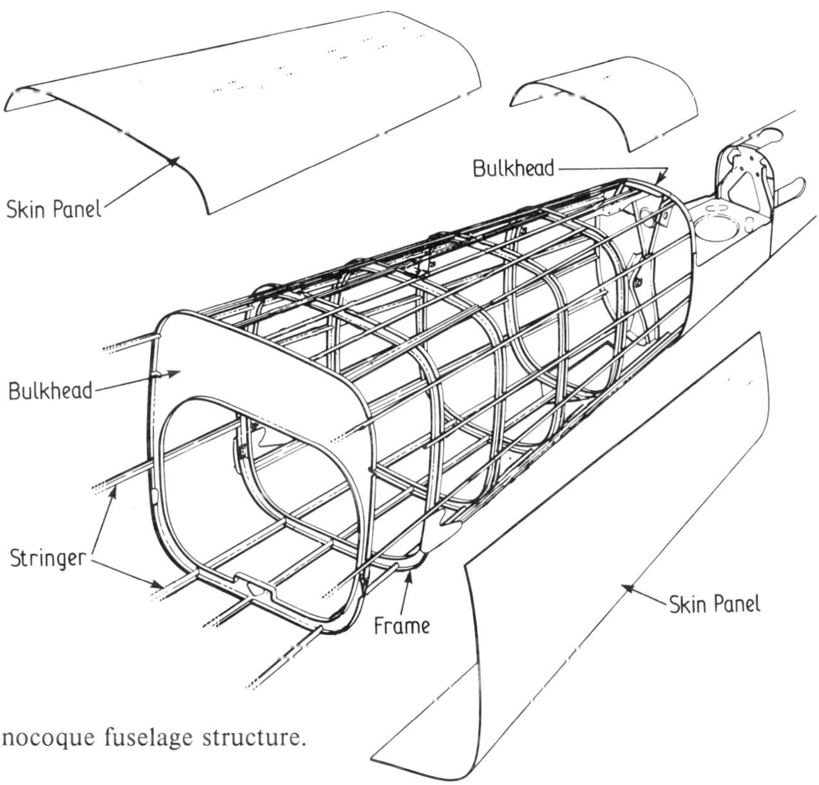

Fig. 3.1
Semi-monocoque fuselage structure.

There are examples where both types of construction are combined in the same aircraft. In the early days of aircraft construction the steel tube fuselage was by far the most favoured among the designers and many examples are still flying. This is also true of earlier light helicopters where steel tubes were used extensively for structural elements. However in the present day the semi-monocoque type of fuselage constructed from sheet metal is the industry standard. In the steel tube fuselage light gauge steel tubes are welded together to form a space frame to which the other parts

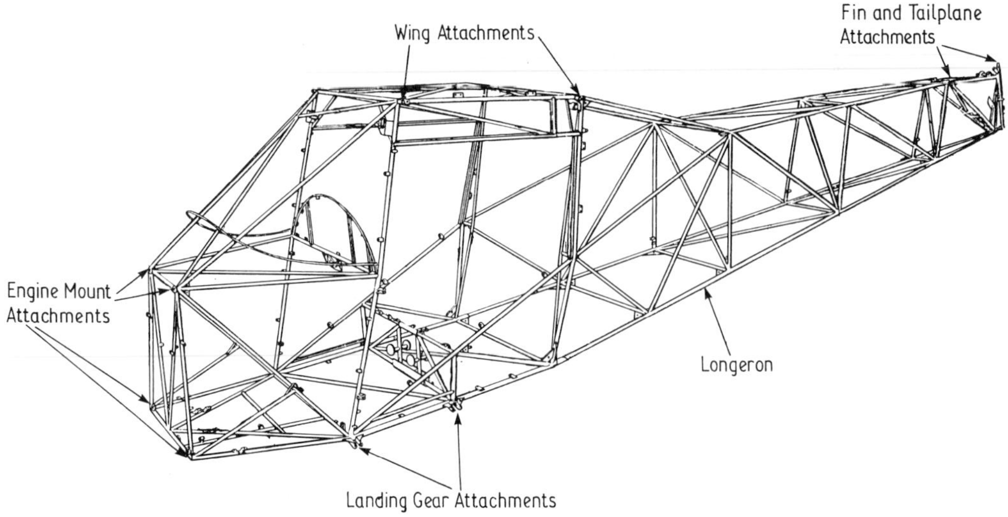

Fig. 3.2 Steel tube fuselage structure.

of the aircraft may be attached. This forms a very strong, reasonably easily constructed and fairly trouble-free basic structure. However the steel tube fuselage requires a large amount of labour-intensive operations to complete it and for this reason, coupled with the fact that a sheet metal fuselage is usually lighter, it has lapsed from the scene for large-scale production.

The semi-monocoque fuselage relies on the skin of the structure to take part of the loads and the skin is stiffened by internal frames, bulkheads and stringers (fig.3.1). Other parts of the aircraft, such as the wings, the landing gear and the engine mounting, are attached to the sheet metal structure using castings or built up bulkheads to form strong points where they are required. Let us spend a moment looking at some of these structural components and their purpose in a typical high-wing, single-engined aircraft. The frames in a semi-monocoque fuselage structure provide the basic shape as well as the strong points already mentioned. For example, two frames, one behind the other, are used to provide the mainplane attachments on each side of the fuselage. The cabin floor is attached to the lower part of the same frames and thus the useful load of the aircraft is connected to the main lifting surfaces. In a similar manner the lower portion of another pair of frames is connected together by castings to provide the attachment points for the main landing gear. At the ends of the aircraft the frames are usually solid and are termed bulkheads. In the USA the term bulkhead is often used for any transverse member whilst the term beam is commonly used for heavy longitudinal members. The engine firewall is a typical bulkhead constructed, like many smaller bulkheads, from sheet metal and standard angle or channel sections. In order to increase the stiffness of the structure between the frames, stringers are riveted in place between them. Stringers may be of rolled section where light loads are involved or of extruded section in order to carry heavier loads. In some light aircraft stringers are not used, their function being achieved by folding the longitudinal edges of the skins through two right angles, thus forming a channel section. This type of construction reduces the number of parts used and also reduces weight.

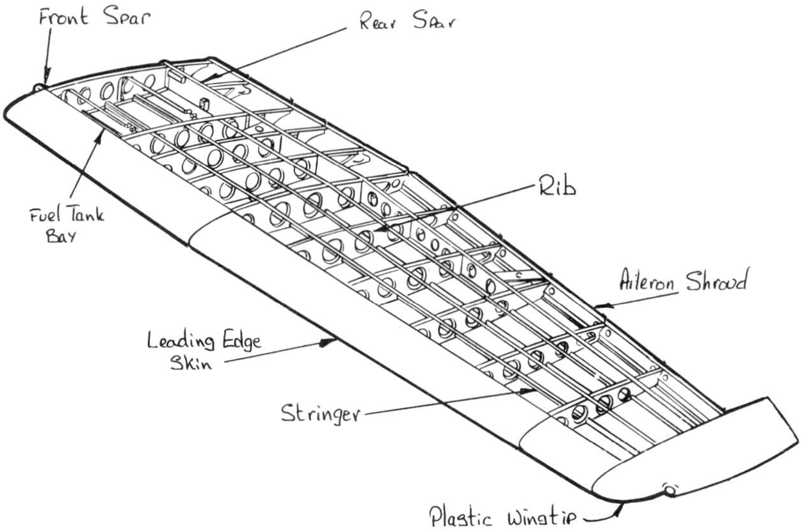

Fig. 3.3 Semi-monocoque wing structure.

A sheet metal wing (fig.3.3) is usually constructed around a foundation of two spars, main and auxiliary. In many cases the main spar is a built-up construction of 'I' section whilst the auxiliary spar is commonly of folded construction. The shape of the wing is determined by ribs to which is attached the sheet metal skin. As in the semi-monocoque fuselage the wing skins are load-bearing and commonly provide all the drag bracing for the wing. The inboard ends of the spar are reinforced with additional sections in order to provide the primary attachments. Outboard of the lift strut attachment the spar is tapered in width and thickness as the bending moment decreases. The auxiliary, or rear, spar provides the attachment points for flaps and ailerons and these are normally positioned in line with main ribs in order to spread the loads through the wing.

It is unusual for a steel tube fuselage to be fitted with sheet metal wings. A fabric covered wing (fig.3.4) is more common, the loads being taken by the ribs and spars. Diagonal drag bracing must be provided for this type of wing.

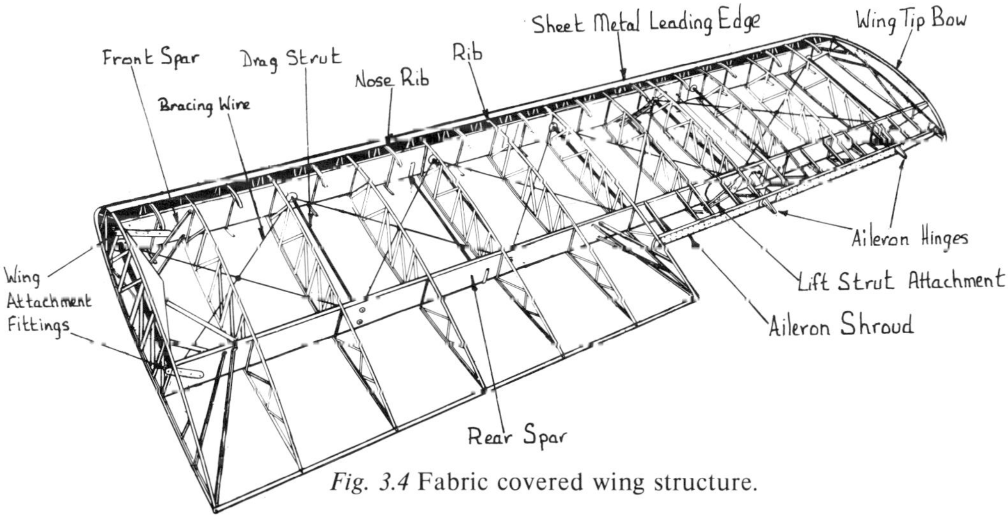

Fig. 3.4 Fabric covered wing structure.

Materials

In a steel tube fuselage the type of tubing used is extremely important if the strength is to be maintained. Not only must it be strong but it must also bc casily weldable by the gas process. For these reasons a typical steel contains alloying elements of chromium and molybdenum. American built aircraft use steel tube conforming to specification SAE 4130. The SAE specification number consists of a group of digits, the first of which represents the type of steel, the second the percentage of the principal alloying element, and the last two or three digits the percentage (in hundredths of 1%) of carbon in the alloy. For example, SAE 4130 indicates a molybdenum steel containing 1% chromium and 0.3% carbon. A steel of this composition has a tensile strength of approximately 40 tons/sq.in. (approximately 620 530 kPa) in the normalised condition. In the U.K. the British Standards Specification T45 has usually been used for structural manufacture. The figure 45 in this particular case stands for 45 tons/sq.in. (approximately 694 992 kPa). T45 has about the same carbon content as SAE 4130 but the principal alloying element is manganese with a small addition of nickel.

Light alloys used in the construction of small aircraft are usually aluminium alloyed with copper with small amounts of manganese and magnesium added. The standard alloy specification used for American light aircraft is SAE 2024 although different tempers of the same alloy may be employed. As in the case of steel specifications the various digits are used to indicate the major alloying elements, the first, 2, in this alloy indicating copper. The second digit, 0, indicates that there is no special control of impurities whilst the last two digits are the same as the superseded designation for this alloy, i.e. 24s. The specification number is followed by the temper, -T3, -T4 and - T42 being the most common. T3 indicates that the material has been cold worked, e.g. by rolling, after solution heat treatment and is the usual temper of sheet stock as supplied from the mill. Such material may be formed into simple parts involving only a single curvature but is unsuitable for more complex forming. To form such parts as channel section frames or double curvature skins the material is first annealed, or softened, this being the - 0 temper. After forming, the part is solution heat treated to the -T4 or -T42 temper. In this case there is no cold working but only natural ageing. The important thing for the engineer to remember is that material treated to -T3 is stronger than material treated to -T4 or -T42. For this reason -T3 may be used to repair -T4 but not vice versa. Light alloy sheet used in aircraft is usually clad on the faces with high purity aluminium. This covering improves the resistance to corrosion but to be effective it must not be scratched or damaged; such material is commonly known as ‘Alclad’.

In the U.K. British Standard Specification L72 is similar to SAE 2024 and is the most commonly used material. L73 may also be found on light aircraft and is of similar composition, but it is approximately 10% stronger and for that reason L72 must not be used as a direct substitute. Whilst L72 and L73 have been the UK industry standard for a good many years they have now been superseded by L163 and L165 respectively, which may be used as direct replacements. Both are clad alloys with the former normally supplied in the T3 temper and the latter in the T6 temper.

When making repairs to aircraft it is essential that the material specification is known and only the correct material used for repairs. Many manufacturers publish diagrams of the aircraft identifying the different materials used. Standard repair schemes will also indicate the correct material to be used.

Damage to the surface of aircraft quality sheet metal can have two effects. In the case of a clad sheet the resistance to corrosion is decreased, as has already been mentioned. However scratches or tool marks on the surface of any material will provide a point of stress concentration which could lead to premature failure.

For these reasons sheets should be carefully stored on edge in a purpose-built wooden rack with each sheet separated from its neighbour by a sheet of heavy paper or card. It is preferable that offcuts of metal be stored separately to reduce the likelihood of scratching larger sheets. Where the material specification is printed right across the sheet identification is no problem. If, however, it is marked on one end only material should always be cut from the other end in order to maintain the identity of the sheet.

Marking out

If care in storing sheet metal is important care is equally important in handling the sheet in marking out and cutting. Absolutely essential is a clean, flat, non-scratch surface that will handle an 8ft. x 4ft. (2 440mm x 1 220mm) sheet, the usual stock size for aircraft material. A sheet of plywood makes a very suitable surface provided it is kept clean and free of grit and swarf. A sheet of insulating board provides a good surface and is cheaper but it will not last as long. The plywood sheet is also suitable as a base when drilling rivet holes in sheet material. A bench is satisfactory for marking out small pieces of material but it is difficult to mark out large sheets unless the plywood is laid on the floor.

It has been emphasised that scratches will reduce the strength of the material so it should come as no surprise that the use of a scriber to mark out lines is prohibited. An HB pencil is quite suitable, provided it is kept sharp. When laying out lines at right angles a square will be necessary with a long steel rule to join up points, as well as to measure. Where a very large square is necessary such an item can be bought as a 'roofing square' from most tool shops. Small circles can be marked out using an ordinary pencil compass but for larger circles a set of trammels, or a beam compass, will be necessary. It is very important to cross-check both dimensions and shape at this stage or an expensive sheet of metal may be ruined. If a square is being marked out do not rely on the use of a try square alone for establishing the angles. Check the accuracy of your work by measuring the diagonals, which should of course be equal. Finally before you cut or drill the metal check the marking once again.

Cutting sheet metal

Sheet metal may be cut with either a guillotine or with tin snips. Where straight edges are required the guillotine is much faster and much more accurate than tin snips. Additionally it will cut much thicker material than is possible by hand. However when it comes to curved lines there is little choice and tin snips come into their own.

A word of warning. Guillotines, even with guards properly fitted, are quite capable of cutting other things than sheet metal, fingers for example. Keep them clear under all circumstances. When using tin snips the cut metal edge is very sharp and is usually passing over the top of the hand operating the snips. Wear a leather glove when carrying out this sort of operation.

When using the guillotine to cut to a line, the sheet is placed on the bed of the machine in front of the cutting blade and the cutting line aligned with the cutting edge on the bed. The sheet is cut by stepping on the treadle whilst the sheet is held securely in place by the hold-down clamp. It is not easy to get the cutting line exactly in line with the cutting edge. An old tinsmith's trick makes it easier. Nick each end of your pencil line with a pair of snips, feed the sheet into the guillotine and pull it back gently until the nicks catch on the cutting edge. A perfect cut every time! A second edge may be cut square to the first by placing the cut edge firmly against the squaring fence at either end of the bed. These are fixed at right angles to the cutting edge.

A large pair of straight tin snips is essential for straight cutting work. These are conventional snips as used by many industries. For most work what are known as 'aviation tinsnips' are superior to conventional snips and shears. Aviation tinsnips are a compound action tool with serrations on one jaw, thus enabling thicker metal to be cut. They are made in left-handed, right-handed and straight versions.

After cutting a sheet it will be necessary to remove the burr formed along the edge. Apart from being a danger to fingers this burr can prevent the sheet seating properly in the joint or can damage the adjacent sheet. Everyone has their own way of removing this burr, some using a file and some a scraper. A suitable tool can be made from a piece of wood rebated to accept a curved-tooth body file (fig.3.5). In use the rebate in the wood runs along the edge of the sheet and the file trims off the burr. The tool should be inclined slightly so that the file does not score the surface of the sheet.

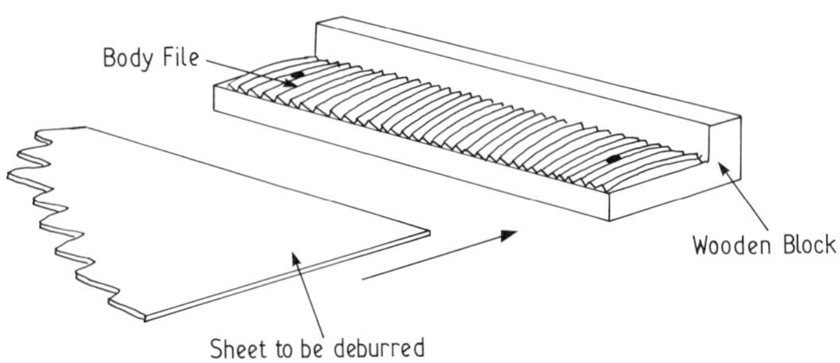

Fig. 3.5 Sheet deburring tool.

One particular job that comes along from time to time is the cutting of additional holes in an instrument panel for additional instruments. This can be done by marking out the hole and drilling a series of holes inside the circumference. The holes can be joined with a pad saw and the job finished with a half-round file. But this is a messy and time-consuming way to do it. A hole saw or an adjustable hole cutter will make a much neater and faster job. If a lot of this type of work is envisaged templates should be made to cover the standard sizes of instruments in order to facilitate marking out.

Drilling sheet metal

With the majority of aircraft using riveted construction the correct drilling of holes assumes major importance. Whilst a hand drill may be used on occasions most work is accomplished using a pneumatic (or 'windy') drill. There are good reasons for the choice of this tool for aircraft work in preference to the familiar electric drill. For one thing the risk of fire is completely removed and so permits work even when fuel vapour is around. A second reason is that an electric drill has a lot of momentum and continues to run for a while after the trigger is released and this makes it more difficult to drill a series of holes reasonably quickly. Thirdly it is smaller than its electric equivalent, which is important in aircraft work. However a pneumatic drill does require an adequate compressor to power it and even the smallest use about 5 cu.ft. of air per minute at 80 lb/sq.in. Many drills use 10 cfm and this must be taken into consideration when buying either a drill or a compressor.

Rivet Diameter	Drill Sizes		Drill Speed RPM	Bucking Bar Weight (lb)
	Pilot	Ream		
3/32″	3/32″ (2.4 mm)	No. 40 (2.5 mm)	4 500	2-3
1/8″	1/8″ (3.2 mm)	No. 30 (3.3 mm)	3 000	3-4
5/32″	5/32″ (4.0 mm)	No. 20 (4.1 mm)	2 500	3-4.5
3/16″	3/16″ (4.8 mm)	No. 10 (4.9 mm)	2 000	4-5
1/4″	1/4″ (6.4 mm)	F (6.5 mm)	1 500	5-6.5
5/16″	5/16″ (7.9 mm)	O (8.0 mm)	1 200	6-7

Fig. 3.6 Rivet tool data.

A further point to which we must give consideration is drill speed. The table (fig.3.6) shows the speeds required for different size drills in various materials. Since common rivet diameters in light aircraft range from 3/32 in. to 5/32 in. we can see that we have a speed range of between approximately 4 500 rpm and 2 500 rpm. Although the higher speed will be more efficient for the smaller drills it will cause the larger ones to overheat, so when making a choice of tool aim at the lower end of the speed range. Drill speed can be controlled by using a pressure regulator, such as is used with a spray gun, but the torque will be reduced as well.

The drilling of aluminium alloys is usually carried out without a lubricant, although a proprietary cutting oil should be used on steels. But in either case the most important factor is a sharp drill. For some reason most fitters will make sure that tools are sharp and in good order, with the exception of the humble twist drill. The small sizes should be scrapped when they are blunt but the larger ones may be resharpened using a proper jig. Grinding a drill freehand is rarely satisfactory unless you are very experienced. If the point is off-centre, or the cutting edges are not equal in length the drill will cut the hole oversize and this cannot be tolerated in aircraft work. A blunt drill tends to make the user apply greater force to the drill and the result is out-of-square holes and broken drills. It really is not worth it.

As with most operations there are one or two safety precautions that should not be ignored. The air exhaust of a pneumatic drill tends to throw the swarf in all directions, including your eyes. So do wear a wide, lightweight pair of goggles to protect your eyes. Fingers can suffer if the piece being drilled is not clamped down properly. As the drill breaks through so the piece of metal will tend to spin and if your fingers are the only clamps they could be badly cut. The problem is more acute if the drill is blunt and with the larger sizes of drill.

Before drilling a hole in any material it is necessary to centre punch at the correct location in order to provide a start for the drill. Thin sheet should always be supported when centre punching but the only really satisfactory tool is the automatic centre punch. With this tool the force of the blow can be adjusted and with even light structures deformation is avoided. When larger holes are being drilled it is important to drill through with a smaller drill first in order to provide a pilot. This will prevent the larger drill wandering and producing an oversize hole! The drill must, of course, be held square to the surface being drilled and kept square whilst it is being withdrawn. If the hole is not square to the surface the fastener cannot seat properly and the strength of the joint will be reduced. If the drill is not withdrawn squarely the hole will be elongated, again weakening the joint. If the drill is at right angles to the surface, the drill and its reflection on the sheet will make a straight line.

During aircraft repair it is often necessary to drill holes in awkward places which are not accessible to the usual equipment. Right-angle drills and flexible drives are available for these situations and can ease what would otherwise be a difficult problem. However bear in mind that there is a stubby series and an extra long series of twist drills that may just permit the hole to be drilled with standard equipment. The smaller sizes of extra long drills will flex slightly and thus permit alignment in difficult circumstances.

In thicker sheets it is sometimes necessary to countersink holes for flush rivets or screw heads. This is achieved using a special countersinking drill with or without a pilot. Use the drill speeds shown in the table in order to avoid chattering and poor cutting. Measure the outside diameter of the countersink cutter and locate this in the left-hand column of fig.3.6 and use the appropriate speed listed in column 4. The piloted cutter is much less prone to chattering but it is necessary to have a different cutter, or at least a different pilot, for each size of hole. Some tools have built-in depth stops but in any case take particular care to have the cutter square to the surface and check the depth frequently. Also make sure that the correct angle cutter is used, for angles from 75° to 120° have been used for different fasteners. The wrong depth or angle of countersink will seriously weaken a structural joint.

The deburring of sheets after cutting to size has been discussed already. It is no less necessary when drilling operations have been carried out. When the hole has passed through several thicknesses of material it will be necessary to dismantle the joint and deburr each hole individually. A suitable tool for this operation is a twist drill larger than the hole but it must only be rotated by hand and just sufficiently to remove the burr. A 5/16 in. (8 mm) drill, the shank of which has been wrapped with tape, will handle most hole sizes used in light aircraft without difficulty. Do make sure that all swarf has been completely removed from the joint before reassembling.

Sometimes it is necessary to drill holes in a new skin to line up with holes in the structure it overlaps. Obviously it is best to drill through from the back (known as 'back drilling') but this is not always possible. Under these circumstances the strap duplicator can be used (fig.3.7). This is made from two old hacksaw blades with the

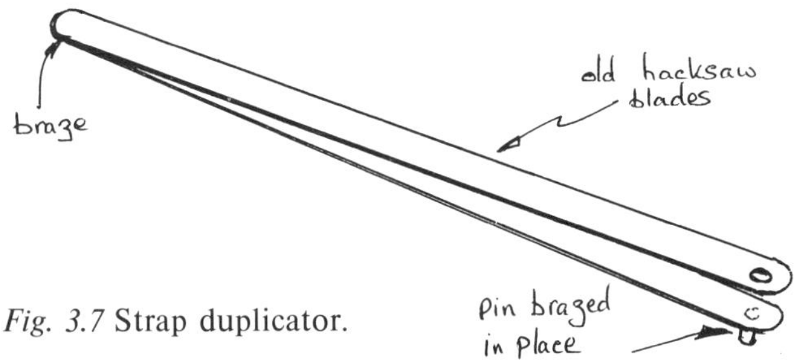

Fig. 3.7 Strap duplicator.

teeth ground off and the two parts brazed together. A pin of a size appropriate to the rivet hole is brazed in one of the remaining holes and must line up exactly with the other hole. In use the pin locates in the blind hole and the drill is started through the hole in the other arm. Once the location has been marked remove the strap duplicator and complete the drilling in the normal way.

Bending sheet metal

The forming of many sheet metal components for modern light aircraft is carried out by using press tools. However there are many parts containing straight bends that are formed on bending tools of one sort or another, particularly during repair and modification work. Doublers fitted under the skin to support radio aerials are normally flanged in order to stiffen the support provided and are a typical example of the type of part an engineer will need to make from time to time.

Before looking at some of the methods that may be employed in bending sheet metal we must first consider various principles that apply to bending operations. When the sheet is produced by rolling at the mill it is left with a 'grain' along the length of the sheet which may be seen by viewing the surface of the sheet against the light. Where possible flanges should be bent across the grain in the sheet where heat-treated material is being used. The reason is simply that the material is less likely to crack along the bend line during forming if the bend is made across the grain. However at times it is required to make bends parallel to the grain although it may be necessary to make the bend over a radius larger than the minimum.

Use of the correct radius for the bend is of paramount importance if cracking is to be avoided. Bend radius is expressed as a function of the thickness of the sheet and should not be less than 3t for -T3 temper material. Annealed material (0 Temper) may be bent with a bend radius of as little as 1t. It must be emphasised that bend radii quoted are *minimum* radii and it may be necessary to use a larger radius as previously mentioned.

In laying out any sheet metal part which has bent or flanged areas, allowance must be calculated for the amount of metal required to make the bends. If this bend allowance is not incorporated when marking out the sheet for forming, the dimensions of the finished part will be incorrect. The bend allowance is dependent upon the material thickness, the radius of bend and the angle of the bend and may be calculated from the empirical formula:

$$\text{Bend Allowance} = N(0.01743R + 0.0078T)$$

where N = number of degrees of bend
R = radius of bend (inside)
T = thickness of material

T and R must, of course, be in the same units.

A slightly simpler, although less accurate formula is,

$$\text{Bend Allowance} = \frac{2\pi R \times N}{360}$$

Where N = number of degrees of bend
R = (inside radius of bend + ½ thickness of sheet).

To simplify the practical application of bend allowance it is usual to use a table in which the allowance has already been calculated. Such a table is given for 90° bends over various radii (fig.3.8). Should the bend be more or less than 90° the allowance

90° INSIDE RADIUS OF BEND.	STANDARD WIRE GAUGE AND EQUIVALENTS IN DECIMAL INCHES.																					
	25	24	23	22	21	20	19	18	17	16	15	14	13	12	11	10	9	8	7	6	5	4
INCHES	·020	·022	·024	·028	·032	·036	·040	·048	·056	·064	·072	·080	·092	·104	·116	·128	·144	·160	·176	·192	·212	·232
3/8	·605	·606	·608	·611	·614	·617	·620	·626	·632	·639	·645	·651	·661	·670	·679	·688	·702	·715	·728	·740	·755	·771
11/32	·556	·557	·559	·562	·565	·568	·571	·577	·583	·590	·596	·603	·612	·621	·631	·640	·653	·666	·678	·691	·706	·722
5/16	·507	·508	·510	·513	·516	·519	·522	·528	·534	·541	·547	·553	·563	·572	·582	·591	·604	·617	·630	·643	·658	·674
9/32	·458	·459	·461	·464	·467	·470	·473	·479	·485	·492	·498	·504	·513	·524	·533	·542	·555	·568	·581	·594	·610	·626
1/4	·408	·410	·412	·415	·418	·421	·425	·430	·436	·443	·449	·456	·465	·474	·484	·493	·506	·519	·532	·545	·560	·576
7/32	·359	·361	·363	·366	·369	·372	·375	·381	·389	·396	·402	·408	·418	·427	·437	·446	·459	·472	·485	·498	·513	·529
3/16	·310	·312	·314	·317	·320	·323	·326	·332	·338	·345	·351	·357	·367	·376	·386	·395	·408	·421	·434			
5/32	·262	·265	·268	·271	·275	·278	·284	·290	·297	·303	·309	·319	·328	·338	·347	·360	·373					
1/8	·212	·214	·216	·219	·222	·225	·228	·234	·240	·247	·253	·259	·269	·278	·288							
7/64	·188	·189	·191	·194	·197	·200	·203	·209	·215	·222	·228	·234	·244	·253								
3/32	·163	·165	·167	·170	·173	·176	·179	·185	·191	·198	·204	·210										
5/64	·139	·140	·142	·145	·148	·151	·154	·160	·166	·173	·179											
1/16	·114	·115	·117	·120	·123	·126	·130	·136	·142													
3/64	·089	·091	·093	·096	·099	·102	·105															
1/32	·065	·066	·071	·071																		

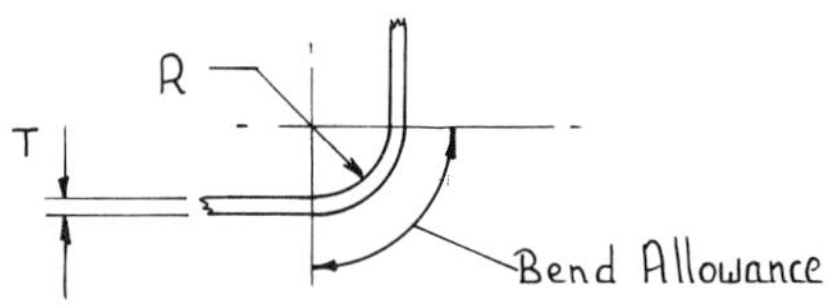

Fig. 3.8 Bend allowance table.

may be calculated by multiplying by the factor $\frac{N}{(90)}$ where N = number of degrees of bend. Due to the natural springiness of the metal it will be necessary to revolve the metal being bent for a greater angle than desired in order to allow for springback. An extra 3° - 5° is a sufficient allowance for commonly used light alloys.

In order to make the bend in the correct place a sighting, or bend, line is necessary which can be lined up with the edge of the bending bar. Such a line is marked out one radius away from one end of the bend as shown in the diagram (fig.3.9).

Wherever two bends intersect material must be removed to prevent strains being set up in the corner that would cause the metal to crack. This is achieved by drilling a relief hole the centre of which is located at the intersection of the bend tangent lines. A general guide for the minimum diameter of a relief hole is three times the thickness of the sheet. However for sheet less than 0.040 in. thick it should not be less than ⅛ in. diameter. Again it must be borne in mind that these are minimum dimensions and should be increased as necessary, even to the extent of doubling the size of the hole.

So far a fair amount has been said about how to bend sheet metal but nothing has been included with regard to the tooling required. The methods used in an aircraft

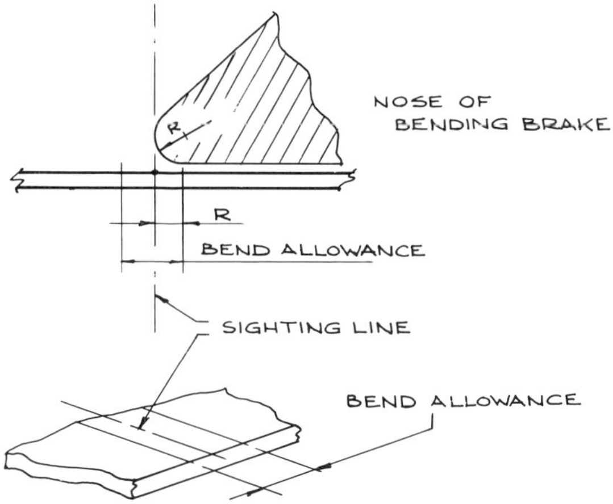

Fig. 3.9 Bend sighting line.

manufacturing facility can differ widely from those used in a workshop specialising in aircraft repair where parts are usually made on a one-off basis. Small items, such as flanged doublers and gussets, may be readily produced using a pair of bending bars clamped in a vice. In use the material is clamped between the bars and is bent over the radiused edge of one of the bars using a hardwood block to ensure that the bend is straight.

Satisfactory as bending bars may be for small flanges they are unsuitable for forming long bends or bends in primary structural parts. Also it is not possible to form a bend beyond 90°. For such work a cornice brake is used and is probably the most common form of bending machine in use.

In use the sheet is clamped in place by the upper leaf. Raising the lower leaf bends the sheet metal round the nose of the upper leaf. Both upper and lower leaves are adjustable for the thickness of the sheet. One advantage of the machine is that the sheet may be slipped through to make a bend in any position. Due to the shape of the upper leaf bends may be made to considerably more than 90° and there is clearance behind the leaf to permit the forming of a channel section. The standard radius on the nose section of a cornice brake may be too small for some work and in any case only the one radius is available. The radius may be increased by bending a strip of metal to form a shoe, the thickness of the metal being chosen to give the radius required.

There is one job that cannot be done on a cornice brake and that is to flange all four sides of a square. A variant of the cornice brake, known as a box folder, has the upper leaf divided into sections. As many sections as are needed for the length of one side are brought into use, the remaining sections being swung out of the way to provide clearance for the previously folded flanges.

Large radii, such as for a leading edge skin, may be formed over a piece of water pipe of appropriate radius. This is made easier if a female die is made from timber and used in conjunction with the water pipe. Two pieces of heavy timber (2 in. x 4 in. minimum) are screwed to a wooden bench, the gap being a little more than the diameter of the pipe. The channel so formed is lined with cardboard. The sheet metal is laid over the channel with the water pipe on top in line with the centre of the desired bend. Driving the water pipe down forms a neat uniform radius with very little tooling expense.

Basic riveting practice

The vast majority of aircraft structural parts are assembled by riveting. Rivets may be conventional solid rivets or blind rivets for use when access from only one side of the job is possible. Consideration will also be given to a variety of special fasteners and the methods of installing them.

Conventional solid rivets are designed to be strong in shear and are not intended for heavy tension loads. However many of the special fasteners that will be studied are equally satisfactory in tension or shear. The material used for solid rivets depends upon the application and the strength required in the design. The most commonly used rivet material in the U.K. is BSS L86 whilst in the U.S. SAE 2117 holds the field. One of the reasons for this popularity is that neither of these materials requires heat treatment by the user before driving the rivet. Since some materials do require such heat treatment it is most important to be able to identify the material from which a given rivet is made. Various methods are adopted for this and are summarised in the table (fig.3.10). It should be noted that L69 has been replaced by L86 although some rivets made from the former may be found in old stocks. The identification marks do not apply to U.K. manufactured metric size rivets.

Rivet Material Table

U.K. Rivets

Specification	Material	Colour	Heat Treatment Required	Identification Mark	
				B.S.	A.S.
L.36	Aluminium	Black	No	I	A
L.37	Al/Cu.	Plain	Yes	7	D
L.58	Al/Mg.	Green	No	8	X
L.69	Al/Cu.	Violet	No	9	–
L.86	Al/Cu.	Violet	No	0	S
DTD.204	Monel	Plain/cadmium	No	–	M
BS.1109	Mild stl.	Cadmium	No	–	–
DTD.367	Tungum	Cadmium	No	– (non-magnetic)	–

Fig. 3.10 Rivet material table.

U.S. Rivets

Specification	Material	Colour	Heat Treatment Required	Part Number Code	Identification Mark
SAE.1100	Aluminium	Plain	No	A	None
2117	Al/Cu.	Plain	No	AD	Dimple
5056	Al/Mg.	Plain	No	B	Cross
2024	Al/Cu/Mg.	Plain	Yes	DD	Double dash
2017	Al/Cu.	Plain	Yes	D	Raised pip
QQ-S-633	Mild stl.	Plain	No	–	Triangle
QQ-W-423	C.R.steel	Plain	No	C	Dash
QQ-N-281	Monel	Plain	No	M	2 Raised pips or 2 Dimples

Fig.3.10 (continued)

Care should be taken in selecting replacement rivets to ensure that not only is the material correct but that the head style is also correct. This is particularly true of countersunk rivets where there are a variety of different angles available varying from 78° to 120°. Use of a rivet with the incorrect head angle will seriously weaken the joint.

U.S. rivets are given a part number which describes the head style, material, diameter and length. An example will show how the code operates.

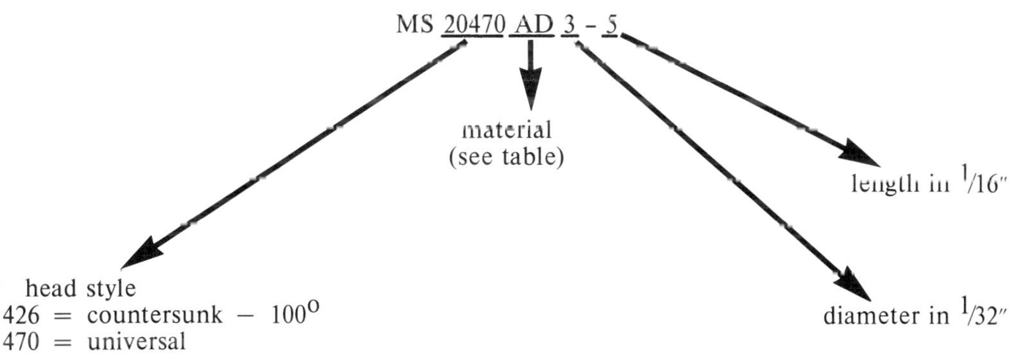

The roundhead and flathead rivets of the old AN series have now largely been replaced by the universal head but, of course, they may still be found on older aircraft. Titanium and stainless steel firewalls are normally riveted with monel rivets in order to prevent corrosion through the action of dissimilar metals. It should be noted that U.K. and U.S. rivets are not directly interchangeble due to differences in strength and therefore repairs should be made with the same type of rivet originally used by the manufacturer.

The riveting process is not particularly difficult but it does normally require two people who have learnt to work together. The method is commonly known as reaction riveting. A suitably shaped set (or snap) is fitted to the pneumatic gun and held on the manufactured head whilst a bucking bar (or dolly) is held against the rivet tail end. When the rivet gun is operated the tail of the rivet is formed into a cheese shape by the reaction of the bucking bar.

Rivet guns vary in size and shape and can have a variety of handles depending on the use for which the gun is intended. Probably the offset handle is the most popular since it permits working right up into a corner. Guns are also classified by the rate of impact, that is the number of blows delivered per minute. For cold riveting of aluminium alloy rivets the gun should operate at a comparatively low rate of impact but still give high energy per blow. Such a gun is often known as a slow-hitting one and delivers up to 2 500 blows per minute. An air regulator mounted on the handle is used to control the amount of air entering the gun and so alters the blow frequency to suit the work in hand. The set is fitted to the gun by unscrewing the retainer spring, inserting the set in the gun barrel and then replacing the spring. Special retaining springs are available for the sets used for countersunk rivets. To adjust the air regulator the set should be held firmly against a block of hardwood and the regulator knob turned, with the trigger pulled, until the correct setting is obtained. This will be gauged by experience. In any case there is no shame in practising on an offcut of sheet metal. Rivet sets are manufactured for each size and head style and only the correct one must be used. *A rivet gun can be dangerous. Treat it like the other kind of gun.*

(i) Never point a rivet gun at anyone at any time.
(ii) Never depress the trigger unless the set is held tightly against a block of wood or a rivet.
(iii) Never leave a rivet gun connected to an air hose when not in use.

Bucking bars vary in style and weight to suit the job in hand. Most sheet metal workers have their own preferences and sometimes it is necessary to make a bar for a certain job, particularly when repairing an assembled aircraft. The weight of the bucking bar is important as too heavy a bar will flatten small rivets and too light a bar will not properly form larger ones. An approximate guide is given in the table (fig.3.6).

When a rivet of the correct material for the job has been chosen it should be pushed through the prepared hole and the protruding length checked. If the length is correct then the rivet will protrude 1.5 times the diameter of the rivet. Rivets may be cut to length using a special rivet cutter which produces a neat, square end to the rivet tail. The rivet is then driven ensuring that the rivet set and bucking bar remain at right angles to the surface of the work. If the rivet has been properly driven the formed head will be 0.5D high and 1.5D in diameter, where D is the diameter of the

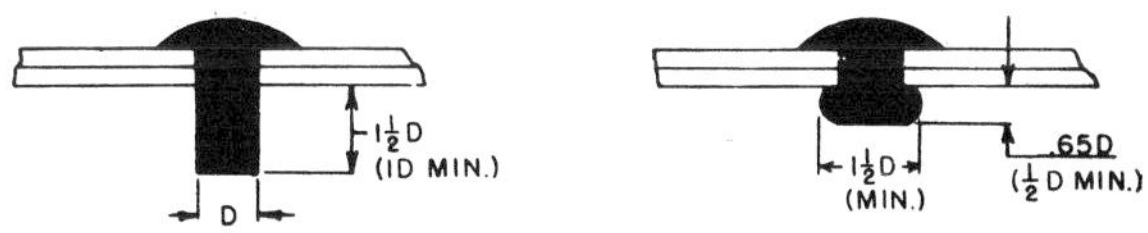

Fig. 3.11 Rivet dimensions.

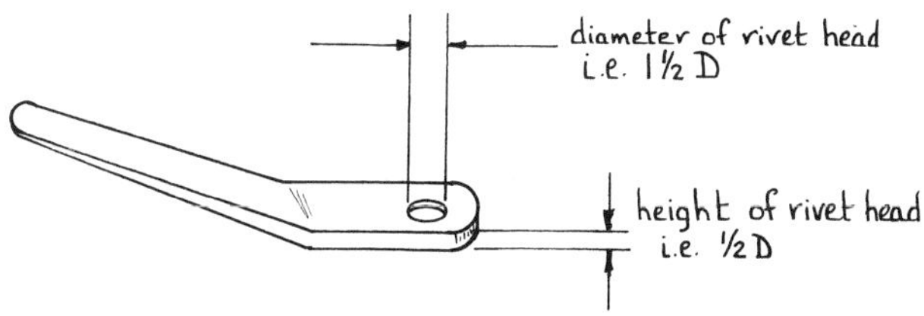

Fig. 3.12 Rivet inspection tool.

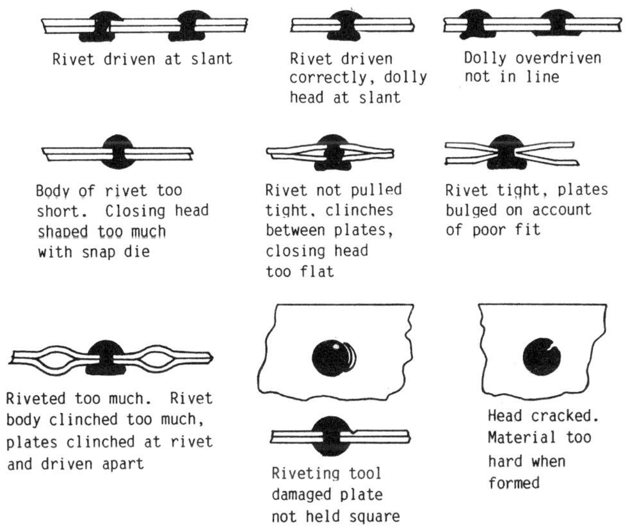

Fig. 3.13 Rivet faults.

rivet (fig.3.11). This can be conveniently checked by using an easily made rivet gauge (fig.3.12). Unfortunately from time to time all does not go well. Various faults, and their causes, are illustrated (fig.3.13). A faulty rivet should not be allowed to remain in place but should be drilled out and a replacement installed. Rivet removal must be done with care if damage to the sheet and hole are to be avoided. Either of these conditions would weaken the joint. First centre punch the manufactured head using an automatic centre punch. Then, using a drill one size smaller than the rivet, drill through the depth of the rivet head. The rivet head will often break away and climb the drill, which is a good signal to withdraw the drill. Failing this insert a pinpunch in the hole and rock it from side to side in order to break off the rivet head. Finally punch out the remaining rivet tail. When punching out the rivet or when centre punching prior to drilling make sure that light structures are adequately supported from the rear by a bucking bar or something similar.

When countersunk rivets are being installed the holes may be prepared in one of two ways. In thick sheet the hole is countersunk using a cutter of the appropriate angle as has been earlier described. However, in thin sheet it would not be possible to

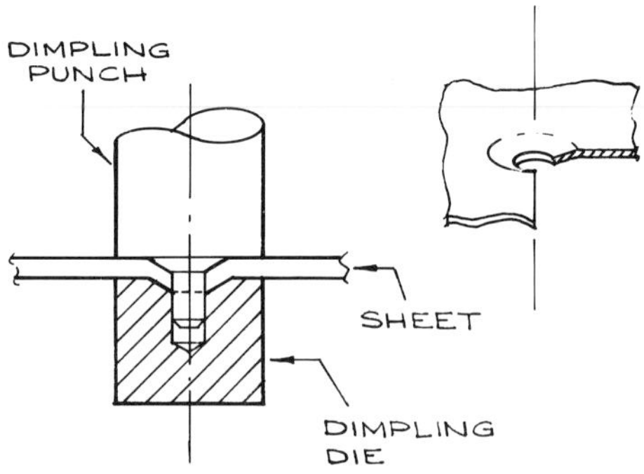

Fig. 3.14 Dimpling.

fully accommodate the depth of the rivet head in the one sheet and, therefore, countersinking would not be satisfactory. In this case the sheets are dimpled or pressed into shape to form the recess for the rivet head.

A male punch and female die are used together in the following way (fig.3.14). A pilot hole, to match the diameter of the pilot on the punch, is drilled through the sheet at the desired location. The tool is assembled through the hole and the die supported on a suitable bucking bar. The punch is then hammered to form the dimple. Only a few blows are necessary and tools are available for hand use or in conjunction with a rivet gun. A countersunk rivet should not be used in place of the punch since it necessitates drilling a larger hole rather than a pilot hole. In the dimpling operation the metal stretches and the hole becomes enlarged. If the hole is full size before dimpling it will be oversize after completing the operation and the joint will have lost some of its strength.

As previously mentioned solid rivets are not the only type of permanent fasteners that are used in aircraft. There are occasions when a fastening must have high strength in both shear and tension and yet be permanent. On other occasions only one side of the job may be accessible and blind rivets are used to solve the problem. They are designed so that they can be installed and expanded from one side of the material.

Probably the best known special fastener is the Hi-Shear Rivet which is frequently used to attach forgings and castings to the airframe structure. Apart from its high strength there is a weight saving of some 65% as compared with making the same joint with nuts and bolts. The Hi-Shear Rivet consists of two parts, an alloy steel pin and an aluminium alloy collar. In use (fig.3.15) the close tolerance pin is fitted in the prepared hole and supported with a suitable bucking bar. The rivet gun is fitted with a special Hi-Shear set and placed over the collar. When driven the work is drawn together and the collar forced into the pin groove. Surplus material from the collar is cut off by the set and ejected through the discharge part. A Hi-Shear Rivet is removed by carefully cutting off the collar with a small chisel and then tapping out the pin with a punch.

A fastener with a similar application to the Hi-Shear rivet, and operating on a similar principle, is the lockbolt. These are marketed under a variety of trade names with minor differences but are very similar to each other. The pin is inserted from one side of the work and the collar slipped over the other end. The special gun exerts a pull which draws the work tightly together, swages the collar in place and then

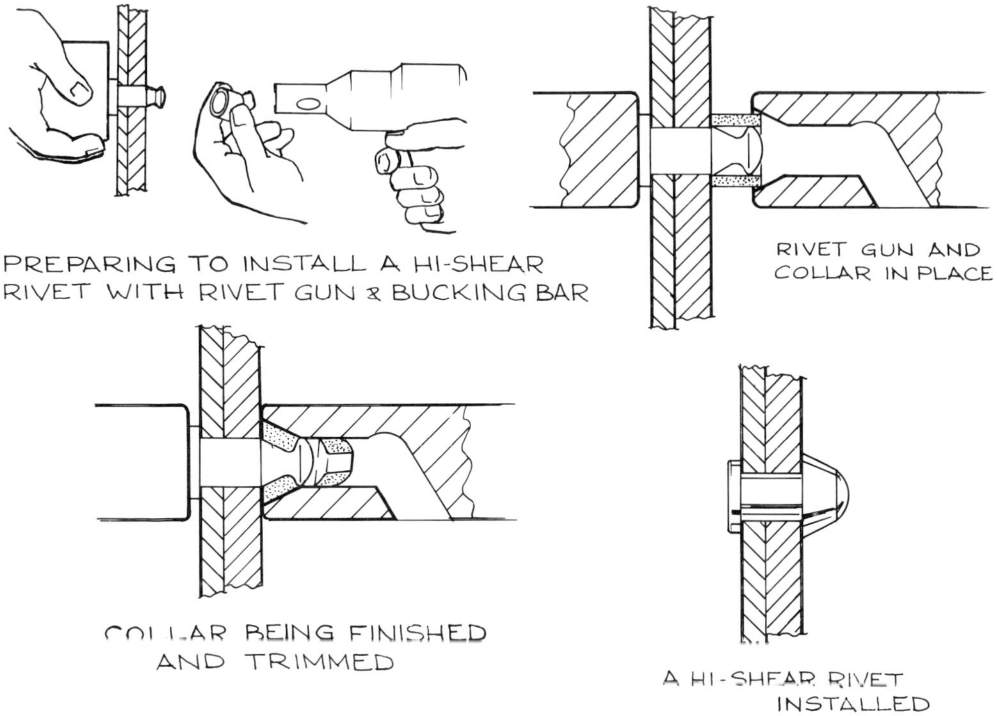

Fig. 3.15 Hi-Shear rivet installation.

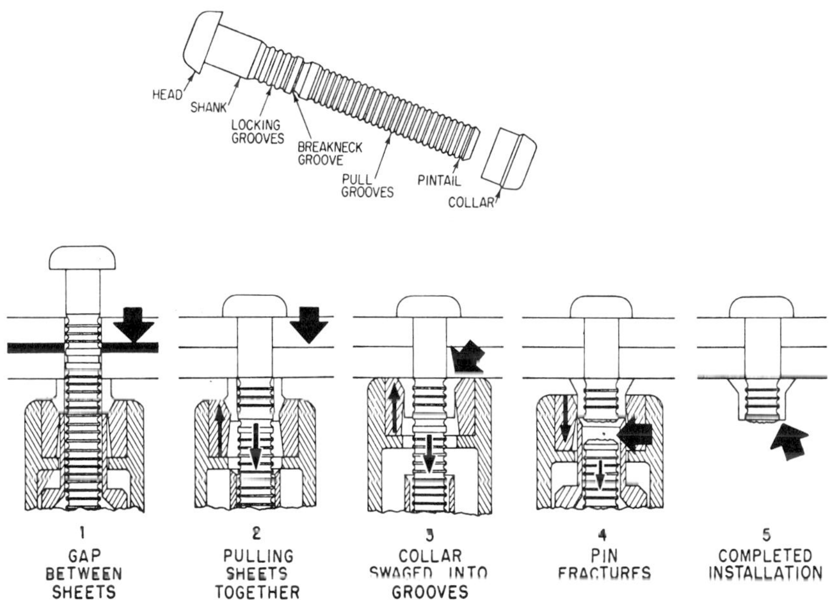

Fig. 3.16 Pull type lockbolt.

fractures the pin flush with the top of the collar (fig.3.16). Removal of a lockbolt is very similar to removal of a Hi-Shear Rivet.

A further type of fastener that is gaining popularity is the Hi-Lok, which comes from the same stable as the Hi-Shear Rivet. The Hi-Lok fastener is unique in that it is basically a high-strength threaded fastener which incorporates the best features of a rivet and a bolt. Additionally it is simpler to install, particularly in areas with restricted access as, for example, in repair work.

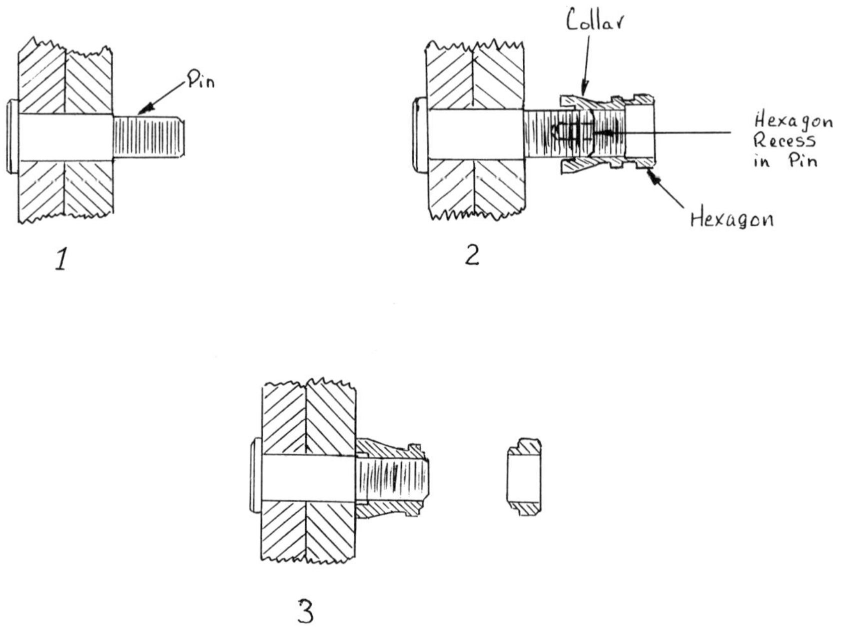

Fig. 3.17 Hi-Lok fastener.

The installation of a Hi-Lok fastener is illustrated in fig.3.17. Insert the pin into the hole prepared in the structure and manually thread the collar on to the pin as far as it will go without using tools. Slip a ring spanner over the hexagon part of the collar and insert an allen key into the hexagon recess in the pin. The collar is now tightened down using the spanner until the upper part of the collar shears off, thus completing the installation. If the pin is an interference fit in the hole it may not be necessary to use the allen key to hold it stationary. The collar may be removed by holding the pin with an allen key and unscrewing the collar using a large pair of pliers with a pipe grip in the jaws.

Now let us examine some of the blind fasteners that can be fitted when access is limited to one side of the work.

The Jo-Bolt is a high-strength steel fastener that can be fitted from one side of the job. The unit is inserted into the hole and the drive tool engaged with the head to hold it stationary (fig.3.18). The core screw of the bolt is rotated by the wrench adapter, causing the sleeve to be expanded over the tapered tail of the nut to form a head on the blind side of the work. At a predetermined torque, the screw fractures at the waisted portion to produce a flush finish with the head on the open side of the work. To remove a Jo-Bolt the head is drilled in much the same way as for a solid rivet and the head is severed by punching the bolt shank right through the material. The American K Bolt is similar in operation.

Leaving aside the high-strength fasteners we come to the range of blind rivets, that is rivets that are designed for the same purpose as ordinary solid rivets but for use

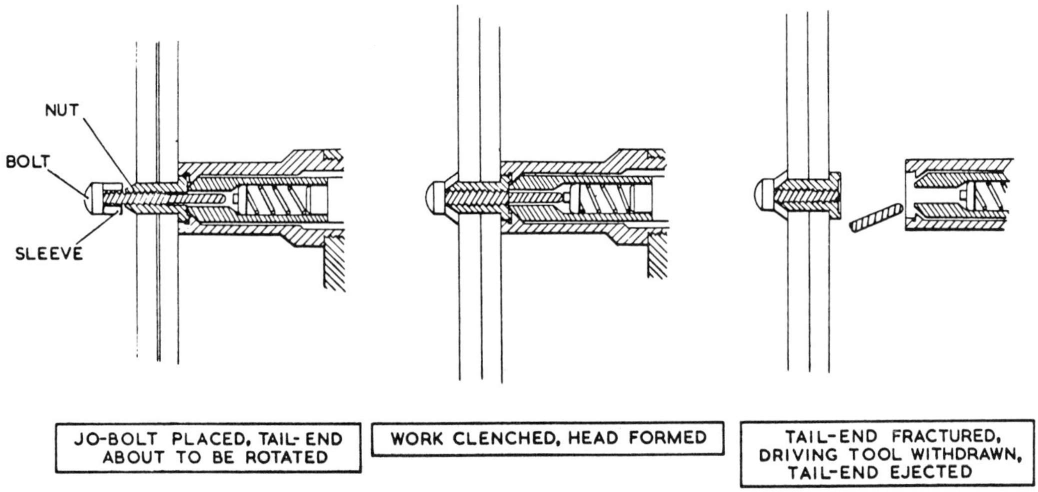

Fig. 3.18 Jo-Bolt installation.

where there is limited access. In this realm the choice is almost unlimited, although certain types are more prevalent than others. One can only make a small selection from among the many varieties available.

Blind rivets fall into two basic categories, friction lock and mechanical lock. Each forms the head by pulling a tapered pin through a hollow rivet and the pin either fractures or is cut off flush. The categorisation is on the basis of the manner in which this pin is retained. In a friction-lock rivet the pin is retained by the interference fit of the pin in the rivet whereas in the mechanical-lock rivet a small collar is swaged into a groove. One point that must be borne in mind, however, is that each make and type must only be installed with its own proper tool.

Having chosen the proper installation tool it is necessary to select the proper grip length rivet using a selector gauge. The stem of the rivet is inserted in the pulling head of the tool, taken to the work and the rivet inserted into the prepared hole (fig.3.19). Taking care to keep the tool at 90° to the surface the tool is operated until the stem breaks. In the case of a friction-lock rivet the stem is then cut off and filed smooth, taking care not to damage the rivet head. This is not necessary with a mechanical-lock rivet, but should the stem be other than flush it could be that the wrong grip length has been used. Blind rivets are not usually as strong as solid rivets and therefore must not be used on a one-for-one replacement basis without increase in diameter. This should not be done without investigation of the shear and bearing strength of the joint to ensure that the original strength is maintained. There is one exception to this rule and that is the bulbed Cherrylock rivet. Removal of blind rivets is very similar to the removal of solid rivets. There are blind rivets where the stem pulls right through and does not break off but these are not often found in structural applications.

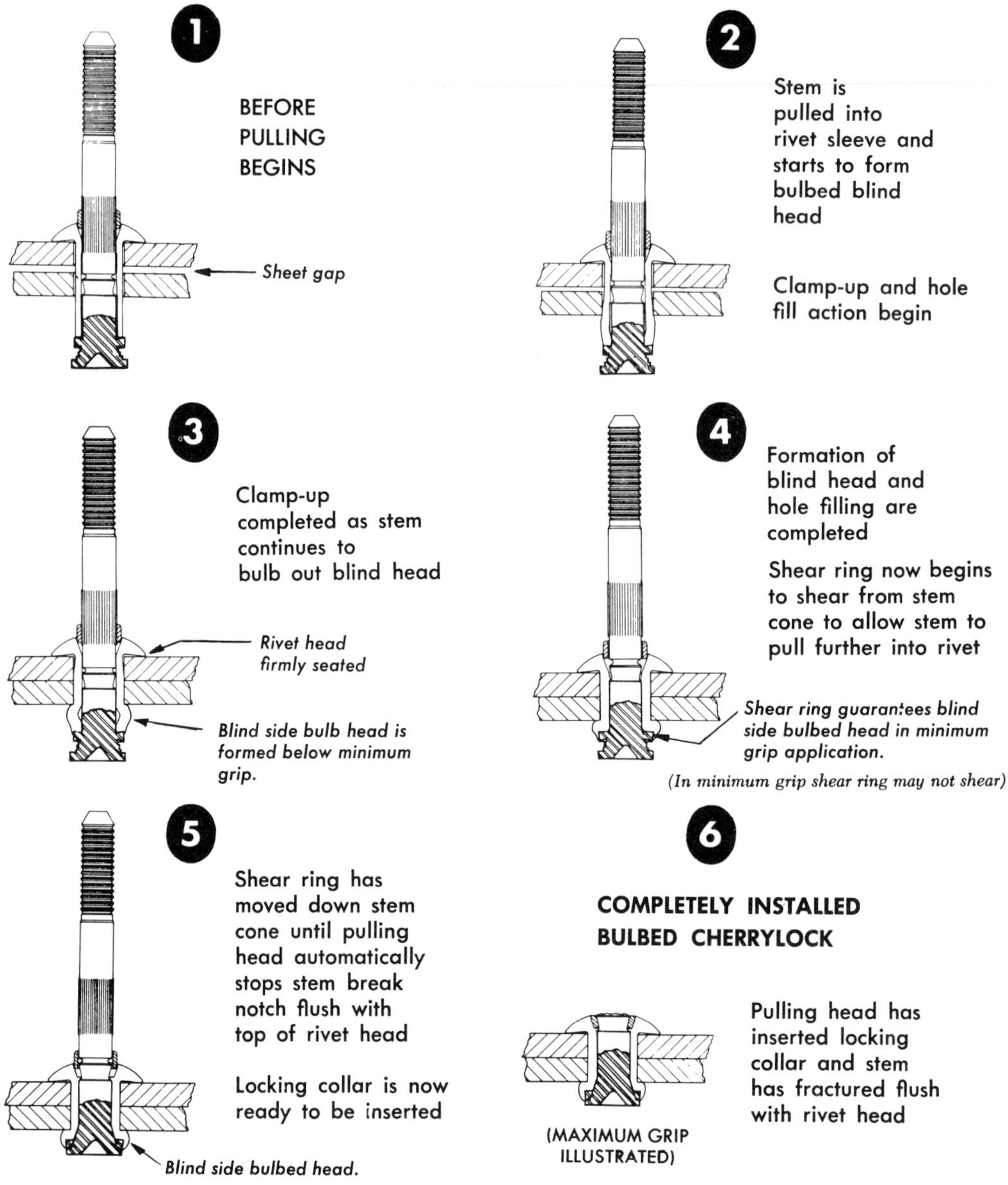

Fig. 3.19 Bulbed Cherrylock blind rivet.

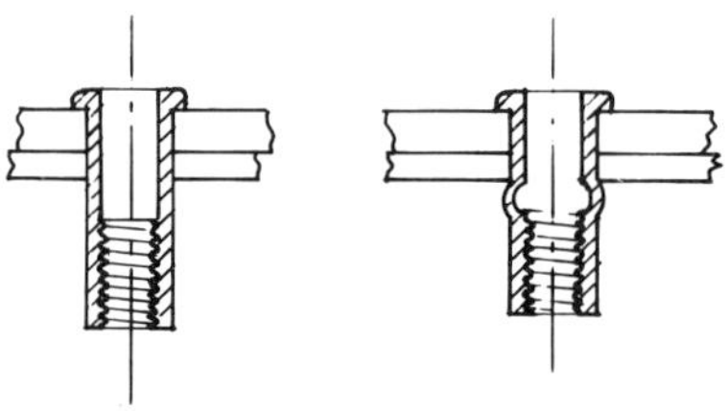

Fig. 3.20 Rivnut.

There is one further type of fastener which is not strictly a blind rivet and yet is installed in much the same way. This is the Rivnut which is used to provide an anchor nut in sheet metal (fig.3.20). The Rivnut is threaded onto a mandrel in the tool, inserted in the hole and the levers squeezed until solid resistance is met. Release the pressure and unscrew the mandrel by pulling the round handle all the way out. Many Rivnuts have an anti-rotation key under the head and a matching keyway must be cut in the hole before installation. Use either a special tool or a needle file.

Assembly of riveted parts

Having mastered the art of riveting it would seem to be easy enough to make a riveted joint, and yet there are a number of points that must be taken into consideration if a satisfactory joint is to be made. Let us assume a simple skin lap joint made along a frame. Having drilled all the holes and deburred both the holes and the sheet there is one further step in preparing the upper sheet in the joint. The sheet should be laid over a bench with the joint edge protruding about ⅛ in. (3 mm) and then dressed down about 0.015 in. (0.4 mm) using a soft-faced hammer. This will ensure a tight fit between the sheets when riveted together. The mating surfaces of the sheets should be thinly smeared with a zinc chromate based jointing compound and the sheets brought together and secured with skin pins. These pins are used to temporarily hold the sheets in correct alignment whilst the riveting proceeds. The screw type of pin (fig.3.21) is infinitely superior where thick skins, curved skins or multiple layers are involved since it will pull the assembly tight and hold it there.

Fig. 3.21 Skin pin – screw type.

The riveting may now proceed. However it is desirable that the operation does not start at one end of the joint and progress along to the other, especially where thin skins are involved. The reason is that thinner skins can stretch slightly as the rivets are driven, resulting in the last few holes being out of line. To avoid this situation rivets should be driven at various locations along the joint starting in the centre and working outwards driving every fourth or fifth rivet. Then the skin pins may be removed and the remaining spaces filled. The final step is to check all the rivets for correct size and shape of the bucked heads and the skins for any accidental damage.

Where large assemblies are being riveted it may be necessary to use jigs or fixtures to ensure that all parts stay in the correct alignment as the work proceeds. In any case the work should be constantly checked for correct alignment using the principles given in Chapter 2. A set of trammels is useful in checking comparable measurements on opposite sides of a structure. For example the diagonal of the door opening on the left side should match the corresponding diagonal on the right side. Always check and recheck. It is too late to change it once it has been riveted.

Repair schemes

Most manufacturers provide basic repair schemes for aircraft of their own make. Sometimes reference will be made to another publication such as 'Acceptable Methods, Techniques, and Practices (AC43.13-1)' published by the Federal Aviation Administration. The manufacturer's recommendations must always be followed since they are precisely matched to his aircraft. Where parts have received major damage the only satisfactory repair is to replace them with new parts from the manufacturer. The repair of tubular structures requires rather more specialised knowledge than can be given in a book of this nature and is usually accomplished by cutting out the damaged tube and making a welded insertion repair. Minor bending may be gently straightened by using G clamps and wood blocks – never by hammering.

Screwed fasteners

A variety of standard screwed fasteners are used in the modern airframe. These vary in strength and application and therefore correct identification is important if mistakes are to be avoided. However a replacement fastener should never be selected by appearance only but by the specific part number quoted in the manufacturer's parts catalogue.

Older British designed aircraft use standard hardware from the AGS (Aircraft General Standards) series of specificiations but in later aircraft these have been superseded by the A series (British Standards – Aircraft) and the AS series (Society of British Aerospace Companies – Aircraft Standards). The later standards contain a much wider range of head styles, high strength materials and lighter weight designs.

American designed aircraft have experienced the same progressive change in specifications for the same reasons. The earlier AN specifications (Airforce/Navy) have largely been superseded by MS (Military Standards) and NAS (National Aerospace Standards) specifications.

In the metric field there has been an updating in design, the current series of threads being designated ISO (International Organisation for Standardisation).

Generally speaking standard screwed fasteners for airframes are divided into bolts, screws and nuts (fig.3.22). However the distinction between bolts and screws varies from one country to another and even from one specification to another. A fairly generally accepted distinction is that a bolt has a portion of plain shank between the head and the thread, known as the grip, whereas the screw is threaded right up to the head. The grip is to provide a smooth bearing portion in a bolted joint and is capable of carrying higher loads. When measuring the length of a bolt it is necessary to distinguish between the grip length and the overall length as shown in the illustrations, since some specifications are based on a nominal grip length and others on a nominal overall length.

Head styles vary considerably according to the standards to which the bolt is manufactured and a number of the more common head styles are illustrated. It is important to be able to distinguish between different designs of head recesses so that the correct driver may be used. Failure to use the correct tool will result in a damaged fastener or an incorrectly torqued screw, or both. This is particularly true of the crosss-point style where there are at least three designs in common use. The twelve-point fastener is turned with a standard twelve-point socket and the internal wrenching bolt with an allen key. Heads and shanks may or may not be drilled for

Fig. 3.22 Standard screwed fasteners.

wire locking or for locking with a cotter pin. Various markings may be used to indicate the material from which the bolt is made but these are by no means universally applied and only the more common markings are shown.

Nuts, like screws and bolts, are also available in a variety of head styles and materials. For example there is little point in employing a high-strength bolt in a tension application if a low-strength nut is used. So it is very important that the right nut is used. Special high-strength washers are also available. Many nuts used in the airframe are self-locking, relying on the deformation of a plastic insert in the crown of the nut to maintain a tight grip on the threads of the bolt or screw. Nuts with a nylon insert are limited to situations where the temperature does not exceed 120°C. Where primary attachments are involved it is becoming common to specify two independent locking devices and this is satisfied by the slotted nut with a nylon insert. When installing this type of nut do make sure that there are no sharp burrs on the cotter pin hole or the nylon insert will be damaged. Self-locking nuts may be re-used provided the self-locking function is not impaired. If, for example, the nut can be screwed fully home on to the bolt using only the fingers then the nut is not acceptable.

There is a further type of screwed fastener that we must consider, one that is commonly known as the self-tapping screw. Occasionally the screw is used to cut its own thread, for example, when attaching some upholstery panels but more often it is used in conjunction with a speed nut. The speed nut is a bowed piece of heat-treated steel strip pierced with a specially shaped hole to allow the screw to engage with it. Various styles are available and are commonly used for unstressed access panels on light aircraft. The pointed screw should be used where it must cut its own thread and the blunt screw when used in conjunction with a speed nut.

Bolted joints

Bolted joints are used in airframe structures where it is necessary to be able to dismantle the parts. Such locations include mainplanes, lift struts, tailplanes, fins, helicopter tailbooms, landing gears and engine mounts. It is extremely important that such bolted joints are correctly assembled in order to avoid fatigue damage through fretting.

Where the joint is a fork joint it is necessary for any play present to be taken up using shims or spacers. If this is not done the lugs of the fork joint will be bent in towards each other when the bolt is tightened and a crack is likely to develop as a result.

Prior to assembly remove all burrs and sharp edges from the holes and fittings and clean out any swarf or foreign matter from the joint area. Usually the surfaces of the joint area are coated with a jointing compound and often one based on zinc chromate is employed. This may be readily distinguished by its yellow or greenish colour. Jointing compound is best obtained in squeeze tubes since this prevents contamination or drying out of the contents. Where a hard setting compound is to be used assembly of the joint must be completed whilst it is still wet in order to avoid uneven stresses in the joint.

Assembling the joint requires some care to ensure that every part is in correct alignment. If the holes are not in exact alignment then the bolt will broach the hole resulting in a poor fit and increased susceptibility to fretting. Alignment whilst the bolts are being driven home can be ensured by the use of tapered pilots (bullets) made from old bolts or from bar stock. In use the pilot is well lubricated and tapped into the hole tapered end first. This is then followed by the bolt, which has first been smeared with jointing compound, and finally the pilot makes its exit from the far side of the joint. In cases where a shouldered bolt is employed in the joint a pilot sleeve will be necessary to blend the thread area to the full diameter of the shank. The sleeve is then removed after installation of the bolt. With the bolt(s) in place the correct washers and nuts may be fitted and tightened to the correct torque value. Where more than one bolt is used the nuts should be tightened progressively, that is, each nut a little at a time, in order to avoid uneven stresses.

Flying control surfaces

Whilst a flying control surface is still a sheet metal structure there are one or two important points that should be borne in mind when making any repairs. The method of construction is usually much lighter than other structural parts and in addition it is statically balanced in order to avoid flutter. Exceptions to the rule of balancing are small trim tabs. Flutter is a rapid vibration that is excited by the airflow over the surface under certain conditions and if allowed to persist can result in the destruction of the control surface.

The methods of repairing control surfaces are the same as for any other sheet metal part but as a rule the repairs should retain the original design of structure as far as possible. By this it is meant that parts should be replaced rather than repaired. Skin splices and small patches are generally acceptable but repairs of a more substantial nature will often result in subsequent difficulty in obtaining the correct static balance. For the same reason priming and painting should be kept to the minimum necessary for adequate protection. When an aircraft is repainted the balance of the flying control surfaces must be checked before the aircraft is released for flight.

The measurement of static balance is not difficult but does require a little care and it must be carried out in a draught-free workshop. The control surface is first supported at its hinge points using two suitable fixtures firmly mounted on a bench (fig.3.23). The balancing beam is placed on the control surface with its centre point directly over the hinge line and the sliding clip adjusted to correctly engage the trailing edge of the control. The beam is now removed without disturbing the clip and balanced as a unit by adding washers to the pin at the end of the beam whilst it is supported at its centre point. The beam is now replaced on the surface and the sliding weight adjusted until balance is achieved, the balance moment being the product of the distance from the hinge line and its actual weight. Generally speaking control surfaces are balanced to have a certain amount of under-balance, that is, they are trailing edge heavy but the exact amount must fall within the limits specified by the manufacturer. Adjustment of the balance is made by adding material to, or removing material from, the balance weights fitted to the surface. Small amounts may be added in the form of standard or 'penny' washers but larger amounts require the addition of lead in an appropriate manner. If a large change is necessary look for the reason and correct the problem at source if at all possible.

Transparencies

Whilst not actually structural parts the transparencies, windows, windscreens and helicopter bubbles are attached to the structure and comprise a part of it. Formed parts are usually made from stretched acrylic but in many aircraft poly-

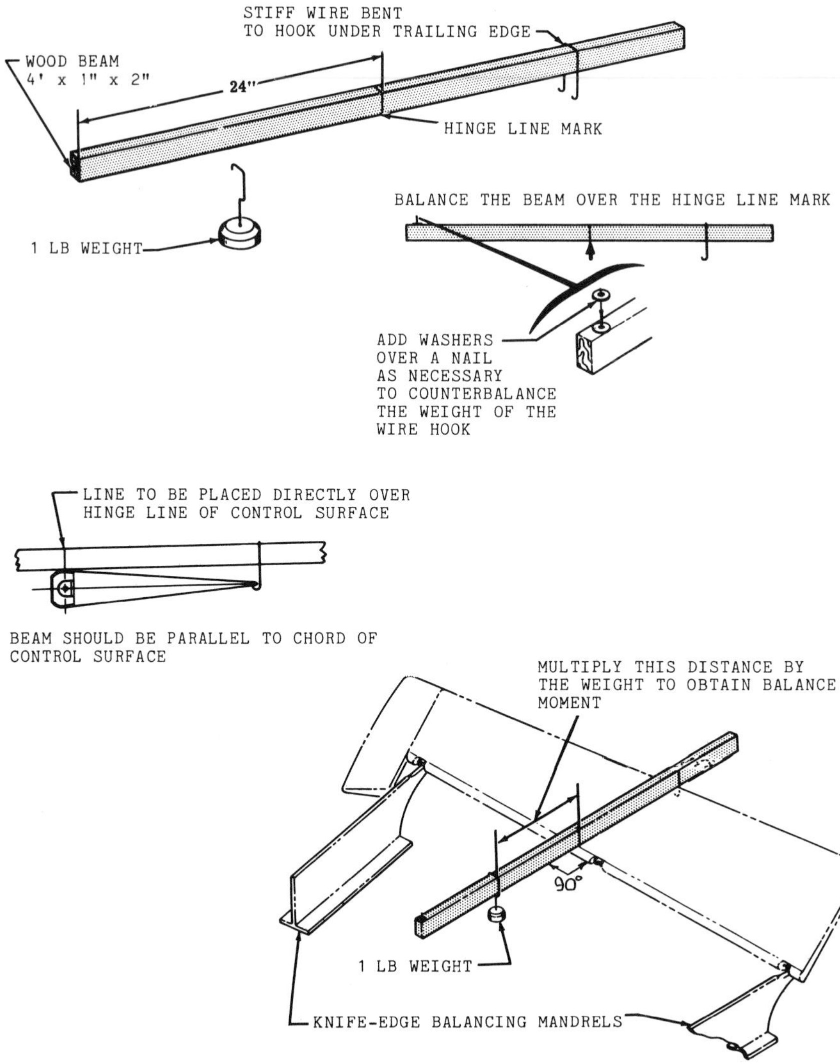

Fig. 3.23 Control surface balance.

methylmethacrylate ('Perspex') is used for flat side windows.

Fitting a new windscreen requires considerable care and should not be attempted alone. The protective paper covering should be left in place except at the edges where a strip about 2 in. (50 mm) wide should be removed to permit trimming and fitting. The first stage in fitting is to offer up the new windscreen to the airframe and check for correct fit. It must fit exactly without any pressure being applied. If it is fitted with stresses remaining in the material cracks will soon develop as a result of vibration and the slight movement that occurs in the airframe structure. The edges should be trimmed to fit exactly using a 30 TPI coping saw, taking care not to twist it at all, finishing the cut edge carefully with a file.

When the fit is correct the holes for the retaining screws may be drilled taking great care, especially as the drill breaks through. The holes should be deburred properly. It should be noted that the holes are normally considerably oversize in order to permit thermal expansion and contraction and in some cases are fitted with rubber bushes. Other windscreens are fitted in a channel or under a retaining strip. The windscreen is sealed into its housing with a suitable non-setting compound with, possibly, a felt strip folded over the edge. The retaining screws and strips may now be fitted making sure that the screws are tightened evenly a little at a time. After fitting, the new installation should be carefully checked for leaks using a water hose. Water that enters will not only make for discomfort of the occupants but will be a corrosion hazard and may enter electronic equipment.

Minor repairs may be made to transparencies provided they do not lie in the pilot's direct line of sight A crack must always be stopdrilled at its end to prevent it spreading with a hole of a minimum of ⅛ in. (3.2 mm) diameter. Drill the hole slightly beyond the end of the crack to ensure that it progresses into the hole and will not bypass it. Stopdrilling alone is satisfactory as a temporary repair but at the next opportunity it should be reinforced with a patch. The patch should be cut from a piece of matching material and glued in place using the correct solvent. It will be necessary to apply pressure to the patch whilst the repair is setting and this can readily be achieved with a small piece of spring steel and two rubber suckers.

Plastic surfaces must never be wiped with a dry cloth since any foreign matter will scratch the surface and, in addition, a static charge will build up that will attract more dirt. Always wash with plenty of clean water. Stubborn patches of dirt may be removed using a mild liquid soap in addition to the water. Plastic surfaces may be polished using one of the proprietary compounds that are available, but these should only be used following a thorough cleaning of the surface. Do not clean or polish surfaces in the direct sun or staining may result.

CHAPTER 4
Fabric Covering

Many light aircraft, especially older designs and gliders, use fabric as the covering for the fuselage and aerofoils. This material has been used from the first days of the aeroplane to provide a light, reasonably strong surface capable of taking the air loads applied to it. The loads carried by a particular piece of fabric are determined by the location on the aircraft, the never-exceed speed and the weight of the aircraft. The highest loads are experienced on the wings due to the lift generated and are usually expressed as a wing loading, i.e. total wing area divided by the maximum weight of the aircraft. These factors will determine the strength of covering required and its method of attachment to the airframe. For this reason it is never permissible to re-cover an aircraft with a lower grade of fabric than was used in the original design.

Materials

Over the years different materials have been used but the modern synthetic fabrics are gradually taking over from the traditional linens and cottons for reasons of both cost and availability. In addition many man-made fabrics are decidedly easier to apply than their natural counterparts and can be lighter by 10% or more. Durability is probably the other factor that tips the scales in favour of the synthetic material, especially in an industrial climate or when used on an agricultural aeroplane. With regard to the longevity of the cover it must be borne in mind that the structure underneath may not be as durable and must be readily accessible for inspection during the life of the fabric. For the same reason a structure must be brought up to first-class condition at the time of re-covering. If the original covering of an aircraft was cotton or linen, changing to a synthetic material, such as Dacron, constitutes a modification and must have the approval of the manufacturer. Where this is not possible, for example where the manufacturer no longer exists, approval must be sought from the airworthiness authority responsible for your aircraft. In Britain this is the Airworthiness Division of the Civil Aviation Authority.

The different specifications are compared in fig.4.1. However some terms used in connection with fabric should first be explained.

Warp	The lengthwise threads in a fabric.
Weft (woof or fill)	The woven crosswise threads in a fabric.
Selvedge	A woven edge to the fabric to prevent the threads unravelling or fraying.
Pinked edge	One cut in a continuous zig-zag to prevent fraying and to provide a longer edge for adhesion.
Thread count	The number of threads per inch in either direction.

Weight — Usually expressed in ounces per square yard.

Strength — Usually expressed in pounds per inch, i.e. the break strength of a strip one inch wide.

Mercerised — Treated with hot caustic soda to shrink and improve strength.

Material	Specification	Min. Tensile Strength-new-undoped - lb/in	Min. Allowed Service Strength undoped - lb/in	Weight oz./sq.yd.	Notes
Grade A mercerised cotton	AMS 3806 (TSO-C15)	80	56	4.0–4.5	For use on all aircraft and required by F.A.A. for aircraft with wing loadings greater than 9 lb./sq.ft. or maximum speeds exceeding 160 m.p.h.
Grade A mercerised cotton	MIL-C-56456	„	„	„	Alternative to AMS 3806.
Intermediate weight mercerised cotton	AMS 3804 (TSO-C14)	65	46	4.0	For use on aircraft with wing loadings less than 9 lb./sq.ft. or maximum speeds not exceeding 160 m.p.h.
Lightweight mercerised cotton	AMS 3802	50	35	3.5	For use on gliders with wing loadings less than 8 lb./sq.ft. or maximum speeds not exceeding 135 m.p.h.
Irish linen (unbleached)	8F1	80	56	4.25	Interchangeable with American Grade A fabric. Better life and rot resistance than Grade A fabric.
Irish linen	DTD 540	90	63		Used originally on De Havilland aircraft.
Scored cotton fabric	DTD 575	50 warp 40 weft	35 28	3.0	Poor resistance to mildew, sunlight or industrial atmosphere. Originally developed for Auster.
Mercerised cotton	5F8	80	56	4.25	Used on DC3 but not the same life span as linen.
Madapolan	F114/343/4A	30	N.A.	2.5	For covering plywood.
Ceconite (polyester)	Ceconite 101 102 103	148 96 N.A.	103 67 N.A.	3.5 2.8 1.7	Heat shrunk Trade name Dacron (For covering plywood)
Razor back (glass fibre)	N.A.	N.A.	N.A.	4.0	Treated for use with butyrate dope only.
Varlene (terylene)	None	200 warp 170 weft	140 119	3.9	

NOTE: Where replacement fabric is stronger than original specification fabric, e.g. when replacing linen with Ceconite, the replacement may remain in service until the minimum figure of the original fabric is reached.

Fig. 4.1 Fabric materials table.

Tools

As with any other job a certain number of basic tools must be available if fabric work is to be properly carried out. A pair of large (10 in./250 mm) scissors with offset handles will be necessary for cutting of fabric and a pair of pinking shears for producing the pinked edge that is required where fabric is joined by doping or cementing. Curved and straight needles for general sewing and long upholsterers' straight needles for rib stitching will also be necessary. A long steel tape is essential for correctly measuring wings and fuselages for the correct amount of fabric. On wooden structures the fabric may be gently held in position with drawing pins or small tacks. Large 'Bulldog' paperclips are very suitable for holding fabric in place on aerofoil trailing edges.

General procedures

Whilst re-covering work is not difficult it must be done under the right conditions. Facilities for re-covering work must be dry, dust free, well ventilated and kept within a temperature range compatible with doping requirements. Barns and damp garages invariably harbour mildew spores, which are easily stirred up into the air and planted on the new fabric. The results can only be described as disastrous. If fabric is stored under damp conditions there is a tendency for it to partially shrink and thus there may not be sufficient shrinkage left to tension the cover on the aircraft. If you are to make up your own envelopes or blankets then a large, smooth, wooden table will be necessary for the marking out and cutting. This would need to be about 7ft x 25ft to cover the largest work but should be large enough for the largest panel. Dirt and oil are the natural enemies of fabric, causing premature deterioration, and should be kept at bay at all times. Good housekeeping will reduce the fire risk by disposing of all oddments of material at the first opportunity. A point to bear in mind is that protective finishes applied to a structure that is to be covered with fabric must be resistant to aircraft dope.

Covering a structure

Covering a wing or fuselage is made easier if a ready-made envelope is bought. This is fabric custom sewn to fit a particular wing or fuselage. The envelope is slipped in place over, for example, the wing and then secured at the root end by sewing or doping. Check on the manufacturer's recommendations for the method of securing the cover. If none are available then keep the old cover and follow the original pattern. If an envelope is not available the wing may be covered by the blanket method. Using a steel tape measure from the trailing edge round the leading edge and back to the trailing edge. Add on sufficient material to make a sewn seam at the trailing edge. Now cut a sufficient number of pieces of this length and lay them side by side until the span of the wing is covered. The pieces should now be machine sewn together with either a French-fell seam or a modified French-fell seam (fig.4.2). If in doubt follow the pattern of the old cover. The blanket is now laid over the wing and located so that the seams fall in between the ribs. Where this is not possible the seam should fall exactly on the rib so that subsequent rib stitching does not pass through the seam. The trailing edges may now be pulled together and the seam closed by hand sewing with either a herringbone (baseball) or overhand stitch (fig.4.3). There should be a minimum of four stitches per inch when sewing. Fitting a cover to a fuselage is

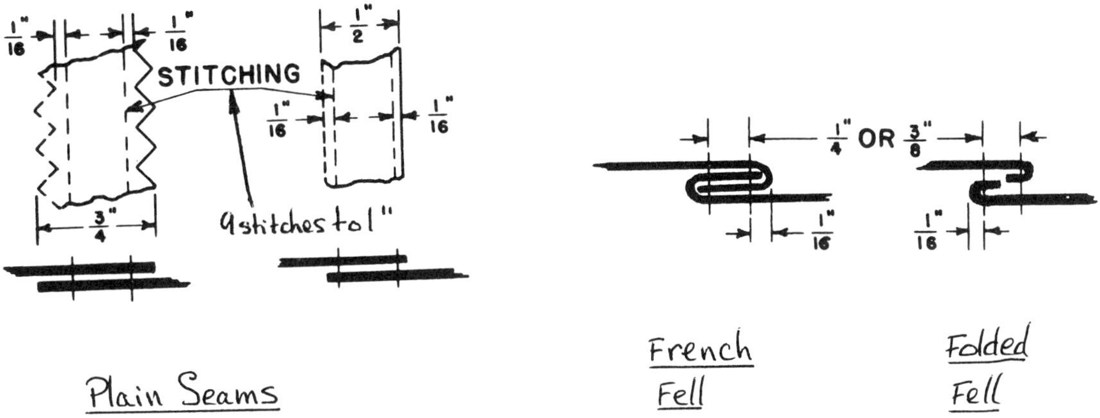

Fig. 4.2 Fabric seams.

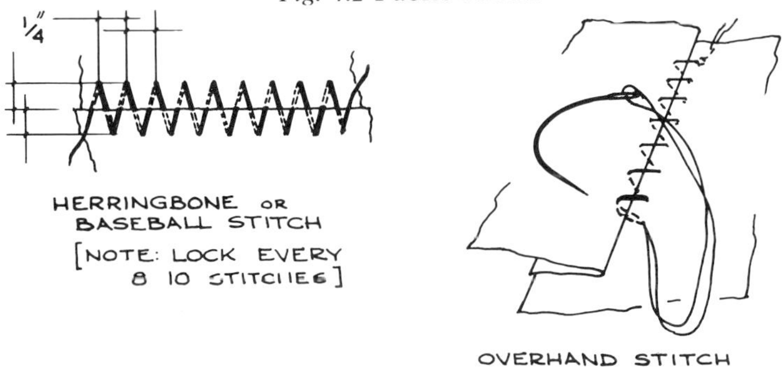

Fig. 4.3 Stitch patterns.

achieved in much the same way as described for a wing. In many cases the fin is covered separately from the fuselage and should be covered first.

Once the cover is located and secured it may be shrunk to remove the slack in the fabric and produce a well-fitting cover. This is done by wetting with water and allowing to dry thoroughly. The cover is now ready for its first coat of dope. Doping is discussed fully in Chapter 5 so the details will not be repeated here.

When the first coat of dope has been completed reinforcing tapes should be applied over the wing ribs by doping. These tapes should be at least as wide as the rib capstrip and in higher speed aircraft (above 250 m.p.h.) the tape should be wide enough to accept the rib stitching. Make sure that the edges are properly doped down. In the case of a wing with a plywood or sheetmetal leading edge support the tapes need only be brought as far as the front spar, but in all other cases they should continue right round the wing.

The fabric must be attached to each wing rib in order to resist the loads applied by the negative pressure above the wing. This is commonly achieved by rib-stitching although screws and washers and wire clips have also been used. In rib-stitching a waxed cord is passed right through the wing close to the rib, round the rib capstrip and back through the wing to be tied off in a knot at the starting point(fig.4.5). It is usual to start at the leading edge of the wing with a double stitch as shown in fig.4.4. The stitching should be continued with single stitches at the same spacing as the original. The spacing in the slipstream area is much closer than that further along the wing, the slipstream area being defined as the diameter of the propeller plus one more rib. Typically for an aircraft with a never-exceed speed of less than 160 m.p.h.

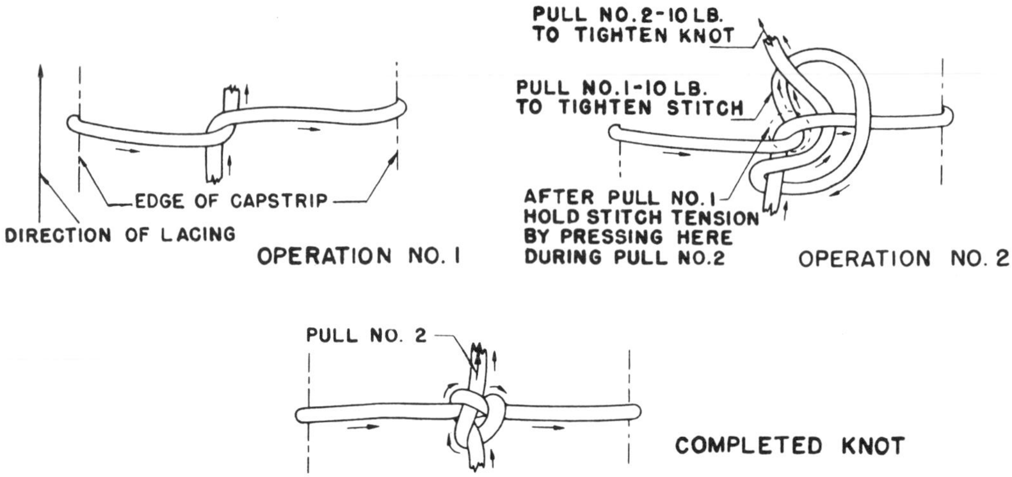

Fig. 4.4 Rib-stitching knot.

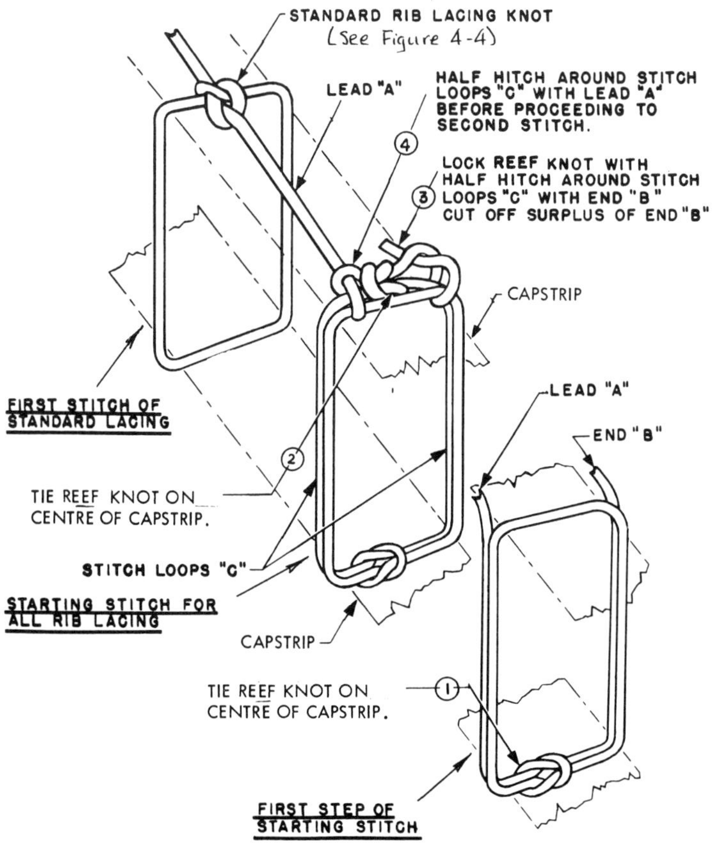

Fig. 4.5 Rib-stitching procedure.

the rib-stitching will be spaced at 2½ in. inside the slipstream area and at 3½ in. out of it. Should it be necessary to join the rib-stitching cord a special knot must be used – the reef knot is unsatisfactory. It is usual to stitch the wing in such a way that the knots appear on the lower surface of a low wing and on the upper surface of a high wing for sake of appearance.

Once the rib-stitching is completed the wing is ready for the finishing surface tape. All sewed seams, lapped edges, rib-stitching and screws must be covered with a 2-in. pinked edge tape doped in place. Where a tape is applied at the trailing edges of aerofoil and control surfaces it should be 3 in. wide. Each edge of the tape should be snipped with a pair of scissors to a depth of 1 in. at intervals of 6 in. In the event of the tape loosening these nicks will allow it to tear off in short lengths without becoming a hazard to the control surfaces. The wing tip bow and the edges of the root rib should also be taped. When taping lay down a wet coat of dope followed immediately by the tape. Press the tape into the dope working out any trapped air and immediately dope over the surface.

Drain grommets should be doped on the lower side of the trailing edges of aerofoil and control surfaces as near to the rib as possible. Grommets should be installed along the centre of the lower fuselage at any point where an internal member could cause water to collect. Do not cut open the holes until all doping is completed, making sure that all fabric is removed. Reinforcing frames for access panels are also doped on at this time.

Covering a component with a synthetic fabric, such as 'Ceconite' (a Dacron material), follows the same pattern as when using natural materials. Panels are sewn in the same way but Dacron sewing thread and rib lacing cord must be used. It is possible to close butt rib and aileron shroud areas of envelopes with a cemented seam, but only the specially approved cement must be used in order to realise the full strength.

The most noticeable difference is when it comes to shrinking the fabric, for 'Ceconite' is shrunk by heat. When the fabric is being attached to the structure it should be left slightly slacker than for cotton fabric. Do not become concerned about fabric wrinkles as they will vanish upon application of heat. A domestic steam iron set to wool (115°C) is the preferred method of applying heat, keeping the iron moving at a speed of about 5 in. per second. The covering should be ironed first from side to side and then from top to bottom to ensure uniform shrinkage. Be careful not to tauten the fabric too much because lightweight structures can be easily deformed. A two-penny piece should just bounce on a correctly tautened covering.

Repairs

One disadvantage of a fabric cover is that it can be easily damaged. However it can be equally easily repaired. A straight or L-shaped tear is sewn with a herringbone stitch and covered with 2-in. pinked edge tape (fig.4.6). Small punctures and tears are

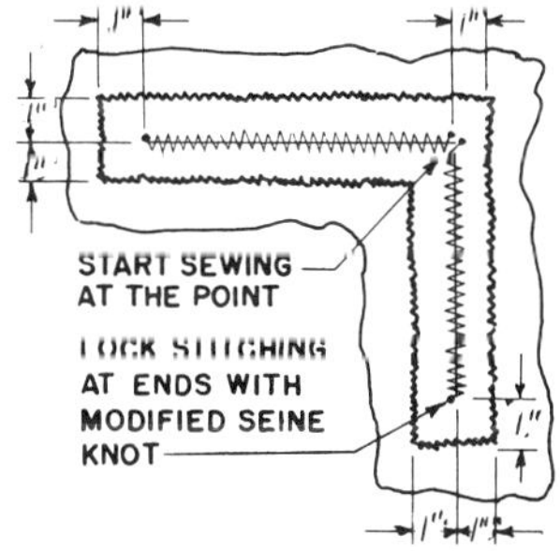

Fig. 4.6 Fabric repair.

simply repaired by doping on a circular pinked edge patch. Larger areas of damage will require a sewn-in patch, doped, and finished with surface tape.

Fabric testing

The testing of fabric for serviceability after some years of usage has been a bone of contention for a long while. Various portable testers are available but their readings are only a guide to the condition of the fabric. They are definitely unreliable on fabric that has been treated with butyrate dope. Laboratory tests require large samples and are not always reliable for used fabric.

There are two methods that are simple and cause a minimum of damage. A firm thumb pressure (about 9 lb/4 kg) whilst the thumb is rotated should neither damage the fabric nor result in the dope loosening. Either is reason for further investigation. A small cut made parallel to the warp or weft with a pocket knife will permit one to insert a little finger. Gentle pulling (2 lb/1 kg) should meet with marked resistance to tearing in line with warp or weft. Should it prove easy to tear the cover in this way the strength of the material is suspect.

If the above tests raise one's suspicions a strip approximately 1.5 in. wide by 6 in. long should be cut parallel to the warp from the suspect area. As much as possible of the dope should be removed, butyrate dope often peeling if it is flexed. The sample should then be washed in a cellulose thinner to remove the remaining dope. A

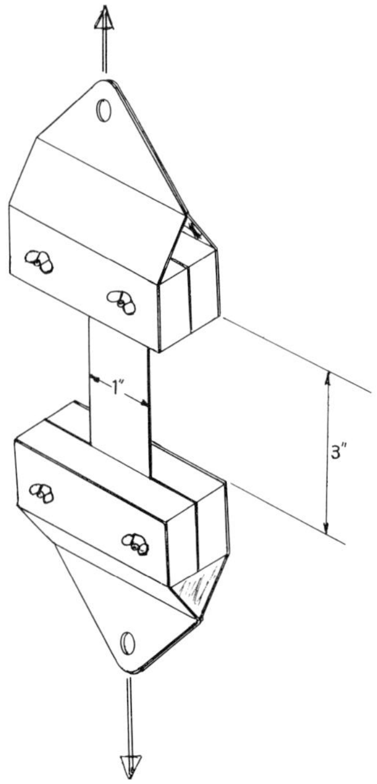

Fig. 4.7 Fabric testing.

simple apparatus for tensile testing may be made as shown in fig.4.7. Threads should be pulled from the long edges of the strip until it is exactly 1 in. wide. It should then be clamped in the jaws of the apparatus so that a 3-in. length remains exposed with the jaws parallel to each other. The apparatus is hung from a firm support and a bucket attached to the lower jaw. Sand is now slowly poured into the bucket until the fabric sample breaks. The bucket, sand and lower jaw are now weighed. If the weight is less than 70% of the new strength per inch of the fabric the cover must be scrapped as unairworthy. Bear in mind that the results will vary with the humidity but that this is also true of other methods.

Deterioration

Deterioration of fabric will take place rapidly if the protective dope finish is allowed to deteriorate. Butyrate finishes, in particular, can be protected by a once a year wax polish to repel moisture, seal out oxygen and clean off dirt. Do not be tempted to use one of the 'quick' waxes but use one of the hard waxes. It is not uncommon for a hangared aircraft to suffer from early fabric deterioration. This condition is usually blamed on poor fabric or dope because it is assumed that an aircraft that is hangared is protected from the enemies of fabric. However this is not true. An unheated dirt-floor hangar may protect the aircraft from wind and sun but there the protection stops, unless the doors are left open. Sun shining on the roof raises the temperature during the day and the warm air absorbs moisture from the ground. The temperature drops at night causing moisture to settle on the aircraft. Atmospheric pressure changes suck damp air into the airframe enclosures. This continuing cycle, with no chance for drying out, eventually soaks the aircraft inside and out, setting the stage for the onset of mildew.

In actual fact if the aircraft is left outside it may not suffer as much. If the doped finish has been well waxed the sun will not attack it. The pigment in the dope and the silver coats will stop ultraviolet light reaching the fabric. But do keep the mice out! They not only chew the fabric but rot it, and the aluminium parts, with their urine.

CHAPTER 5
Surface Finishing

Aircraft finishes have come a long way since the early days of the aeroplane when fabric and structure lives were short. The high capital cost of an aircraft today means that it must be effectively protected to ensure a long and safe life in service. But an aircraft finish is not only important for protection but also for decoration. A properly finished aircraft is a joy to look at and will have a higher resale value. When the aircraft leaves the factory it has been given a finish that is both decorative and protective. It is the responsibility of the maintenance technicians to see that this finish is maintained in such a way that it will keep its beauty and continue this protection.

Safety precautions

Safety precautions always seem to smack of government regulations and whilst such regulations are applicable to aircraft finishing the precautions are more important in this field than perhaps any other. The materials that are used in aircraft finishing can be hazardous due to the very high fire risk as well as being a hazard to personal health. As a result no apology is offered for spending some time on the subject.

From the point of view of flammability nitrate dope is the material that is most likely to ignite. One of the most common causes of ignition is a spark produced by the discharge of static electricity. If a wing is being doped using nitrate dope the fumes from the solvents and the nitrocellulose fill the inside of the wing. When the dope has dried subsequent dry sanding and dusting will build up a static charge on the surface in much the same way as when you slide across a plastic car seat cover. If the operator is wearing rubber-soled shoes he will be at the same electrical potential as the wing and nothing will happen. Should this charge now be lost through touching some metal part of the hangar and he now touches, say, the control cables in the wing then the static charge jumps to the cable creating a spark and igniting the fumes. The best way to prevent this kind of problem is to eliminate the static charge altogether by grounding the surface that is being worked on. A fairly heavy wire connected to the surface and a clean metal part of the hangar will do the job satisfactorily. Clothing that is made from synthetic fibres will build up a static charge much more easily than that made from cotton, and leather-soled shoes will allow any charge to be dissipated to ground. When spraying nitrate dope make sure that the aircraft, the spray gun and the operator are all grounded together.

A further warning is necessary in relation to the standard of housekeeping in the spray shop. Dried nitrate overspray is highly explosive and sweeping will create enough static electricity to ignite it. To clean the floor, swill with water and then sweep whilst it is still wet. If the floor of the spray shop has dried nitrate or butyrate dope overspray on it make sure that it is removed by wet-sweeping before allowing any zinc chromate overspray to mix with it. If these oversprays are mixed spontaneous combustion can result. The overspray from certain enamels can catch fire by themselves if swept up and put in a pail of water. Rags and sponges that have been used to apply one of the phosphoric acid conversion coatings, such as Alodine, are another source of danger from spontaneous combustion if the material is allowed to dry in the rags.

The spraying area must have a sufficient movement of air through it in order to remove the vapours and fumes from the solvents employed in the finishing processes. These fumes are hazardous to health as well as being a fire-risk. Polyurethane paints present special hazards in their applications and the instructions provided by the manufacturers must always be followed. Anyone with cardiac ailments or respiratory disease should not be allowed to spray polyurethanes on a production basis. At the first sign of any irritation of the skin or eyes, difficulty in breathing or a dry cough they should stop work and seek medical advice. Fresh air respirators, gloves and well-fitting overalls are essential when using this material.

Working conditions

Many of the factors that have been mentioned in the previous section are related to the working conditions that are required in a spray shop. However there are factors other than those related to safety that are necessary if a satisfactory finish is to be produced. The air flow through the shop must be high enough to remove all dangerous fumes but it must not be such that it will chill the work in hand and cause such problems as blushing in a dope finish. An explosion-proof fan should be mounted near the floor level since the solvent fumes are heavier than air. The desirable plan is for the air inlet to be in an adjacent room where the air can be brought up to the temperature required for the process in hand. This arrangement will prevent a direct draught upon the work. For doping the temperature should be between 70° and 80°F (21° - 26°C) with the humidity between 20% and 70%. The successful application of a wash primer requires a minimum level of moisture in the air and other materials have their own special requirements. Always check the manufacturer's recommendations and stay with them.

The spray gun

The main tool of the aircraft finisher is the spray gun which seems a simple enough piece of apparatus. However many of the faults in a finish are caused by improper adjustment or use of the spray gun. Basically there are two main types of spray gun in common use, the suction-cup gun and the pressure-fed gun. In the former the container for the material is attached to the gun and the material is sucked out of it and in the latter a much larger remote container is pressurised with air to feed the fluid to the gun. The pressure-fed gun is preferred where complete repainting of an aircraft is required whilst the suction-cup type is more suitable for the smaller job. Nevertheless perfectly good work is carried out on larger jobs with the suction-cup gun provided extra care is taken.

Both types of guns have three valves to give the operator full control of the material that he is spraying (fig.5.1). The air valve is opened when the trigger is first pulled and

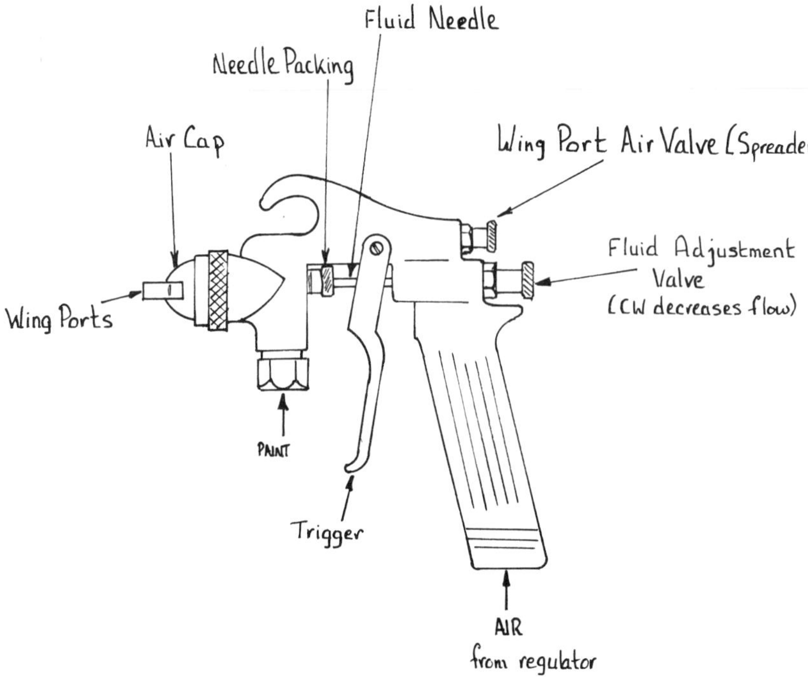

Fig. 5.1 The spray gun.

this allows air to flow out of both the atomising holes and the wing port nozzles. The atomising air is controlled by adjusting the pressure at the regulator but the spray width adjustment (spreader valve) controls the air flow to the wing ports. Continued pulling of the trigger opens the fluid valve so that the material can either be forced out by the pressure in the pressure pot or sucked out by the low pressure developed at the fluid nozzle by the atomising air flow. Adjusting the fluid valve controls the movement of this valve and therefore the amount of material delivered to the nozzle. As the fluid leaves the nozzle it is broken up into very small droplets and sprayed on to the surface being finished.

The wing ports in the air cap are used to control the spray pattern. If no air is allowed to flow from these ports the spray pattern will be circular. As the spreader valve is turned anticlockwise the valve is opened, increasing the air flow and flattening the spray pattern (fig.5.2). As the width of the pattern is increased so the

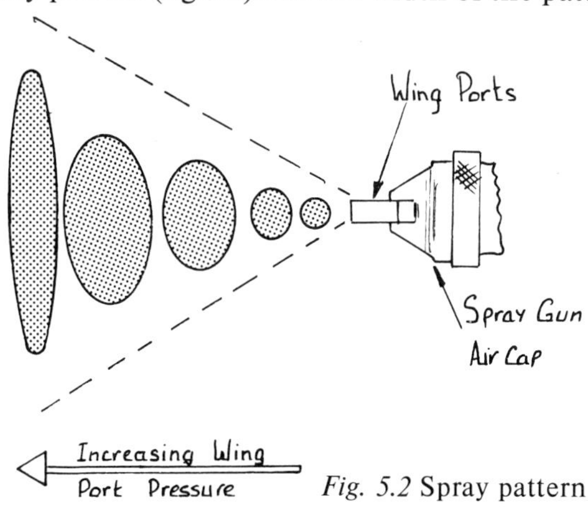

Fig. 5.2 Spray pattern control.

fluid valve must be opened proportionally in order to maintain a constant thickness of paint film. The actual air pressures used at the gun and on the pressure pot will vary with the type of gun used and the material being sprayed. The pressure at the gun should be just enough to atomise the fluid properly and the pressure at the pot just sufficient to adequately feed the gun. The likelihood of such defects as runs, sags, pinholes and orange peel is much reduced by the use of low pressures.

Cleanliness and a little preventive maintenance are the two keys to faultless spray gun performance. However, faults do develop and it is important to be able to recognise and correct them. Spitting, or interruption of the fluid flow, is caused by air entering the fluid passages of the gun. With a suction gun this may be caused by excessive tilt of the gun when the fluid level is low. A dried-out fluid needle packing will also cause this problem and may be usually remedied by lubricating with a light oil. A deteriorated gasket between the gun and the cup is another possible cause. The majority of other operating problems result in a distorted spray pattern and are best

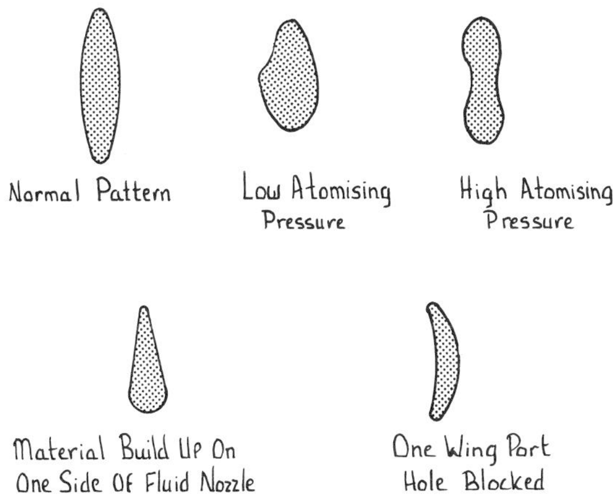

Fig. 5.3 Spray pattern defects.

considered in relation to the shape of the distorted pattern (fig.5.3). The normal pattern is fan shaped, the width of the pattern being dependent upon the adjustment of the spreader valve. The coverage should be even within the area of the pattern. Should there be an excess of material in the centre it could be due to insufficient atomising air pressure and the regulator should be adjusted to compensate. Too high an atomising air pressure will cause the opposite effect, that is the pattern will be waisted or dumb-bell shaped. It may also occur if an attempt is made to obtain too wide a pattern with thin material, in which case the air flow from the wing ports should be reduced and the fluid flow increased. Irregularly shaped patterns indicate that one or more of the air holes in the air cap have become blocked. In such a case the spray gun should be dismantled and completely cleaned.

Immediately work has been completed the spray gun and its equipment must be thoroughly cleaned. Any material remaining in the suction cup should be emptied and the cup cleaned. Partially fill the cup with thinners and spray it until the thinner comes out without any trace of the material. Dry the gun with compressed air. When cleaning a pressure-fed gun first loosen the air cap on the gun and the lid of the pressure pot. Force the fluid in the gun and hose back into the pot by placing a rag over the holes in the air cap and operating the trigger. Clean the pressure pot, partially fill it with thinners and spray through the gun to clean it. The nozzle of the gun may be soaked in thinners for further cleaning, provided that the packings are kept clear of the thinners. Soaking the packings in thinners will harden them and make them useless.

There is no mystique in the use of a spray gun but a fair amount of skill is required which can only be acquired by practice. Once the gun has been correctly adjusted for the material being sprayed the final factor in the quality of a paint job is the application technique. In use the nozzle of the gun should be held at right angles to the surface being sprayed and about 6-8 in. (15-20 cm) from it. The distance is readily judged by the span of an open hand. When spraying, move the gun parallel to the surface keeping it at right angles and at a constant distance. Begin the stroke, then pull the trigger. Release the trigger before completing the stroke. A common fault is to swing the spray gun in an arc instead of keeping it square to the surface which

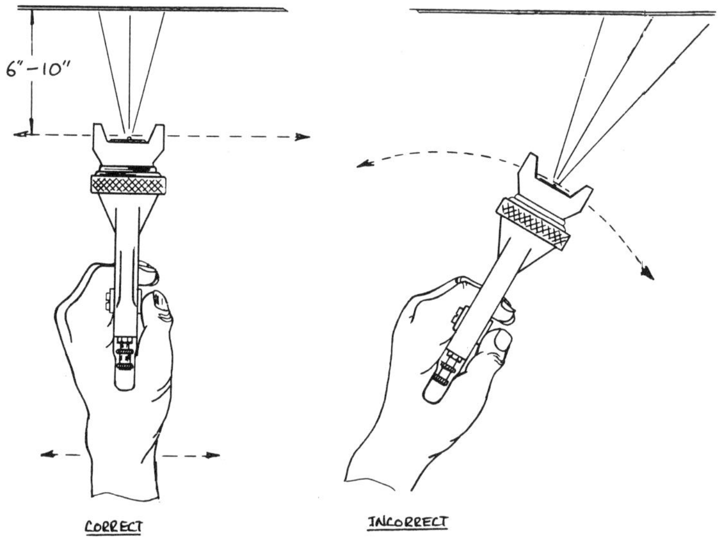

Fig. 5.4 Spray gun handling.

results in an uneven coating (fig.5.4). When painting a part the edges and corners should be sprayed before the flat areas. This is done by spraying along the corner and

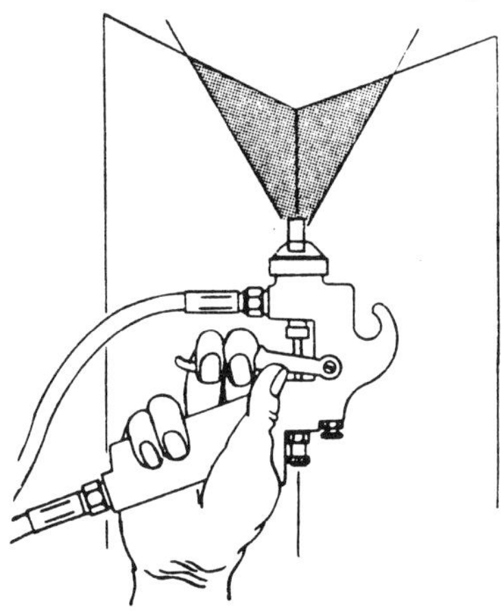

Fig. 5.5 Painting a corner.

gives the thickest coat along the edge, blending out in the flat portion (fig.5.5).

Finishing fabric covered components

Mention has already been made of butyrate and nitrate dopes but a further word of explanation is in order. Butyrate (cellulose acetate butyrate or CAB) dope has become popular in more recent years because it remains more flexible with age, is more fire retardant and absorbs less moisture when compared to nitrate dope. However, butyrate dope does not have the same ability to encapsulate, or completely surround, each fibre in the fabric as does nitrate dope and the latter can be used with advantage for the first coat. Butyrate dope may then be used for the subsequent coats. At no time may nitrate dope be applied over butyrate due to the different order of the solvents used. Incidentally, where butyrate dope is applied over nitrate dope much of its fire retardant property will be lost although the other advantages remain. Never use old dope since it turns acidic with age.

The first coat of dope should be brushed into the fabric so that it penetrates the weave and encapsulates the fibres. To facilitate this the dope should be thinned by up to 50% making sure that it is not so thin as to drip on the inside of the fabric. When doping natural fabrics a fungicide should be used in the first coat in order to reduce the risk of attack by mildew. After the tapes have been applied (see Chapter 4) a further brush coat (of clear butyrate this time) should be applied, thinned sufficiently so that the dope will not 'rope' or pull. 'Roping' is a condition that occurs when the surface of the dope dries when it is being brushed, producing a stringy, uneven surface. This may also be caused by the temperature of the dope being too low for satisfactory application and the addition of thinners will not improve the situation.

After the second coat of dope has completely dried the nap, that is the short fibre ends in a natural fabric, will be standing up giving a rough surface. Very lightly sand this smooth using dry sandpaper and taking the essential precautions already detailed. Follow this with spraying on two full wet cross coats of clear butyrate dope, thinned enough to spray properly. A cross coat is a double coat of paint sprayed on in one direction and then, before it dries to the touch ('flashes off'), at right angles to it. The amount of thinning necessary will depend on the equipment being used but will vary from about 1 part of thinner to 5 parts of dope to equal parts of dope and thinner.

Ultra-violet radiation will damage both fabric and the undercoats of clear dope so a coat of aluminium dope is applied for protection. If a ready-mixed dope is not being used it is prepared by mixing clear dope with aluminium paste in a ratio of about 1 lb of paste to 5 gallons of dope. The dope should be added to the paste rather than the reverse in order to ensure proper mixing and then thinned as necessary for spraying. Lightly sand the clear dope undercoats with 400 grit paper and wipe the surface with a cloth slightly damped with thinner (a 'tack rag'). Spray on one full-bodied cross coat of aluminium dope. After the coat has thoroughly dried wet sand it just enough to produce a smooth surface. A full coat must remain after sanding if it is to protect the fabric and this may be checked by shining a light inside the surface when any thin spots will show up. It must be lightproof. The aluminium surface must now be washed, scrubbed if necessary, in order to remove any traces of loose aluminium powder. Failure to do this thoroughly will result in the finish coats peeling from the aluminium coat.

The aluminium coat is now followed by the finish coats. Spray on a good cross coat of the required colour butyrate dope, the gloss of the final finish being determined by the smoothness of the surface. Such a finish described here gives an adequate balance between quality and cost and will give adequate tautening and protection with a minimum of effort. Many owners and engineers prefer a multi-coat, hand-rubbed finish which then becomes a labour of love. To produce such a finish use as many thin wet coats of butyrate dope as may be necessary, wet sanding with 600 grit paper between each coat. Allow the finish to dry for at least a month, preferably longer if the weather is at all cold or wet. The surface may then be finally smoothed with rubbing compound and waxed.

Dacron fabrics, such as Ceconite, may be finished in much the same way as natural fabrics. The main difficulty is in bonding the first coat of dope to the fabric since the man-made fibres are not able to soak up the dope. In order for the dope to adhere properly the fibres must be completely encapsulated and this is achieved by a careful brush coat of nitrate dope worked into the weave. The first coat should form a wet film on the inside of the fabric, taking care that it does not drip through to the opposite side. Since Dacron is not a natural organic fibre a fungicide is not necessary in the first coat of dope. There are proprietary products supplied by the manufacturers of the fabric which may be used instead of nitrate dope to provide a superior bond to the fabric.

A normal dope finish on Dacron fabrics will show the weave of the fabric due to the pronounced depth of the weave. The addition of more coats simply adds weight without improving the finish. A special high-solids butyrate dope containing aluminium is available and should be used straight after the prime coats when a smooth finish is desired.

Finishing metal components

Whilst the durability of the finish on an aircraft will vary with the conditions of usage and climate no finish will last long if the metal surface has not been properly prepared. The metal must be completely clean, roughened to provide a key for the finish and primed to provide a suitable surface for the top coats. If the metal has already been painted the surface must be restored or stripped before proceeding with any other paintwork.

If the old paint film is basically sound it may be sanded to remove all surface dirt, wax, lubricants and other foreign matter and to roughen it to provide a key for the new paint film. Following the sanding it should be wiped over with a tack rag. In all operations make sure that acrylic window panels and fibreglass and plastic parts are properly masked or the solvents used will attack the plastic and cause deterioration.

Where a paint film is unsound it must be completely stripped. The wax type of paint stripper is normally used, the wax holding the active agent (often methylene chloride) against the surface until it has done its job. Brush on a heavy coat and wait until all the paint in the area has softened and wrinkled. This may then be removed using a non-metallic scraper. If an area dries before it has properly lifted the paint film brush on some more and allow it to remain until the action is complete. Acrylic lacquer will not wrinkle or bubble but it will soften to the point where it may be scraped off. The paint residues must be washed off with water using a bristle brush in the corners to remove stubborn remains. Since the stripper uses a wax it is vital to remove every trace of this by washing the entire surface with a good solvent such as acetone or methyl ethyl ketone (MEK). Toluol or xylol make satisfactory and cheaper substitutes. Lacquer thinner is not suitable as it will not absorb the wax but only spread it around to cause trouble later.

A word of caution. When using paint stripper always wear goggles and rubber gloves. If any is splashed on the skin wash it off immediately with water. If any paint stripper should come into contact with the eye repeatedly flood it with cold water and see a doctor immediately.

If corrosion is found after the surface has been cleaned then assess the nature of the damage and remove the corrosion or replace the part (see Chapter 1).

With a suitably prepared metal base a conversion coating should be applied in order to prevent the recurrence of any corrosion and to microscopically roughen the metal, thus providing a good bond for the primer. A conversion coating is a phosphate film formed on the surface of the metal by the reaction of phosphoric acid with the metal. The acid content is actually very low and after the liquid has remained on the surface for a period of time a thorough flushing with water is sufficient to remove any unconverted acid.

After the surface has been properly prepared and pretreated it is ready for the application of a primer in order to provide a bond with the top coats. There are three types of primer in common use: zinc chromate, wash and epoxy primers.

The most familiar in the aircraft industry is, without doubt, zinc chromate primer in either its yellow or green form. As far as corrosion protection is concerned it is still good but there are other primers which have far better adhesion. Zinc chromate primer is readily sprayed after thinning with a proprietary thinner or toluol. However it should not be used over wash primer nor should it be used as a base for acrylic lacquer. Zinc chromate provides its inhibiting qualities by actually allowing a small quantity of moisture to enter the film, but if there is a small amount of free phosphoric acid in the phosphate film the two will combine and attack the metal. The solvents used in acrylic lacquers will lift a zinc chromate primer unless it has dried for some days.

Wash primer must be mixed prior to use since it uses an acid hardener which provides a conversion coating. Its greatest advantage is the speed with which the top coats may follow. The curing time necessary to allow the proper conversion of the acid is about half an hour and the first top coat should be applied after this time but before eight hours have elapsed. If this is not possible the primer will form a very hard glaze that will make for poor adhesion of the top coat. Should this happen a second coat of primer should be sprayed on as before, but omitting the acid. Correct application of wash primer will result in a very thin film, one thin enough to give a slight amber tint to the aluminium. The most critical aspect of the application of wash primers is the necessity of having sufficient moisture in the air to properly convert the acid into a phosphate film. Proper conversion requires a minimum of 0.09 lb of water per pound of dry air. This is measured by the use of wet and dry bulb thermometers and the chart shown in fig.5.6. Always be very sure to follow the manufacturer's instruction for mixing.

When a polyurethane finish is to be used, or when maximum corrosion protection is required, an epoxy primer is the right choice. By contrast with wash primer this type of primer must be left for at least five hours, and preferably overnight, before applying a top coat of acrylic or enamel. However a polyurethane top coat can be applied after a delay of one hour. The application of top coats should not be left beyond 24 hours or the surface will be too hard to provide a proper bond. If this situation does arise it will be necessary to break the glaze with 600 grit sandpaper before top coating. Epoxy primer requires mixing with a catalyst, following the manufacturer's instructions carefully, and using within six hours of mixing. As with

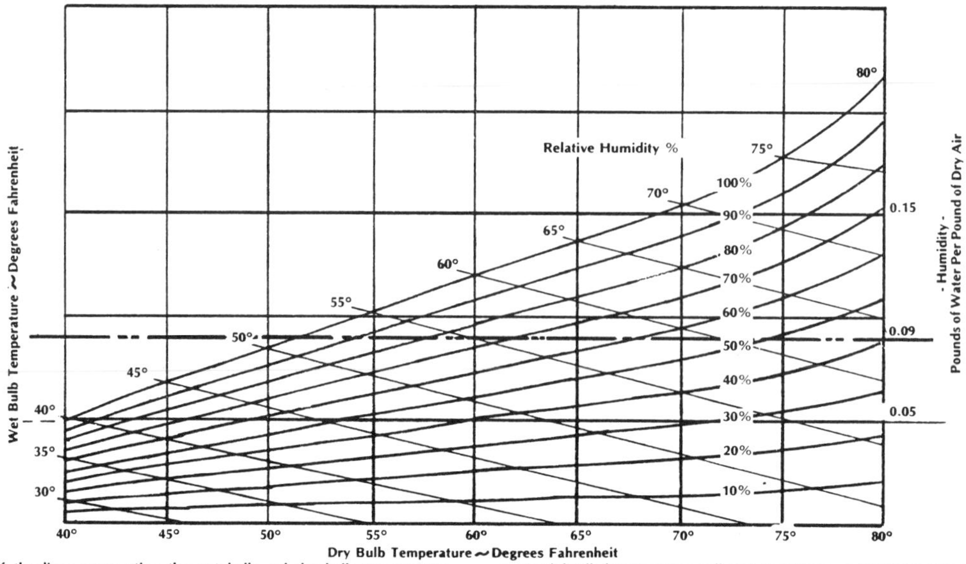

If the lines representing the wet bulb and dry bulb temperatures intersect above the 0.09 pound line, there is sufficient water in the air. If the intersection is between the 0.09 and 0.05 pound lines, you may add one ounce of distilled water to one gallon of the thinner. If the intersection is below the 0.05 pound line, a maximum of two ounces of distilled water may be added to each gallon of thinner to aid the cure of the primer.

Fig. 5.6 Wet and dry bulb chart.

wash primer only a thin coat is necessary, just thick enough to slightly colour the metal.

Top coat materials may be divided into enamels, lacquers and polyurethanes. To differentiate, a lacquer dries by the evaporation of its solvents whereas an enamel cures by the conversion of some of its solvents by heat, oxidation or some other means. A conventional enamel has a solids content of some 45-50% and is thinned, or reduced, by 1 part thinner to 10 parts enamel for normal spray use. The thinning range may vary from about 1 part thinner to 20 parts enamel to about 1 part thinner to 5 parts enamel. In use a light mist coat of enamel is sprayed and left to dry for 15 minutes. This is then followed with a full wet cross coat and allowed to dry for at least 48 hours before masking or taping for the trim.

Acrylic lacquer has gained popularity with the increase of high volume production. The solids content is low by comparison with enamel, being about 30%. Whatever the final colour being applied a white base coat should be applied in order to obtain the correct colour required. The low viscosity and poor binding qualities of the acrylic lacquers tend to lead to the application of coats that are too heavy. They should be thinned about 5 parts thinner to 4 parts lacquer. Spray on a very light mist coat and then follow with at least three cross coats allowing about half an hour drying time between coats. If the material is too heavy, pinholes, or orange peel are likely to show in the finish. The pinholes, or air bubbles, may be so fine as to produce a haze on the finish. This is known as air impingement haze and is caused by the excessive atomising air pressure needed to spray on excessively thick material. The key to good finishes with acrylic lacquer lies in multiple thin coats.

Polyurethane enamel produces a hard, smooth, glossy surface that is highly resistant to chemical attack and therefore finds application where Skydrol hydraulic fluids are used and for agricultural aircraft. In use polyurethane enamel is mixed with a catalyst and allowed to stand for about 15 minutes. It may then be thinned to the proper viscosity for spraying, checking this with a viscosimeter. Time the flow from the cup to the *first* break in flow and check this with the enamel manufacturer's specification. Do take the proper safety precautions when spraying polyurethane enamel. Spray a very light mist coat first and leave it for about fifteen minutes. This should be followed with a full cross coat, being very careful not to get it too thick. A practical 'rule-of-thumb' is to spray until you feel that one more pass will be right - but leave it there. Do not spray that extra pass! The characteristics of the enamel mean that it will flow for at least an hour after application and will cover those spots that appeared thin. Resist the temptation of 'one more pass'. Masking tape may be applied after 24 hours, but the trim should be sprayed straight away and the tape removed as soon as possible. If the tape is left on the surface for a day or so it will be almost impossible to remove.

Whenever an aircraft has been painted it is essential for the static balance of the control surfaces to be checked (see Chapter 3).

Aircraft painting sequence

Any job needs proper planning and a paint job is no exception. The following order of painting will minimise overspray problems.

1. Paint the ends and leading edges of the ailerons and flaps, then the flap and aileron shrouds, the wing tips and the leading and trailing edges.
2. Spray the landing gear, the wheel wells, control horns and hinges.
3. Paint the lower tail surfaces starting at the root and moving outboard, spraying chordwise.
4. Work up the lower side of the fuselage allowing the spray to go up the sides. Work right up to the engine cowlings.
5. Spray the bottom of each wing starting from the root and working towards the tip, spraying chordwise.
6. Spray the fin and move forward painting the top and sides of the fuselage, spraying across the fuselage.
7. Spray the upper sides of the tail surfaces.
8. Paint the upper sides of the wings.

The tack coat should be sprayed at right angles to the prime coat. The final coat should be sprayed in the same direction as the prime coat.

Finishing problems

Various problems have been mentioned in the context of each type of material. However, a summary of common problems is given at the end of this chapter. Colour matching can be a problem even when using a factory-mixed colour and especially when using metallic paints. There are one or two things that are within the control of the operator that will affect the colour match. Proper mixing is extremely important and can be achieved in the following way. Pour off half the contents of the tin into a clean tin and stir thoroughly with a paddle until *every bit* of pigment is in suspension. Repeat the stirring for the other half tin. Pour the contents of the first tin into the second and inspect it to make sure that every trace of pigment has been loosened from the bottom of the tin. If all is well thoroughly mix the paint by pouring it from one tin to the other as many times as is necessary. Colour match will also be affected by the atomising air pressure, the amount of thinner used and the number of coats applied.

Painting over a red has been a long-standing problem for the red pigment is soluble in the solvents used in topcoats and will bleed right through whatever is sprayed on top. New pigments have been developed but be careful when refinishing or repairing an existing paint scheme.

SUMMARY OF COMMON FINISHING PROBLEMS

ROUGH FINISH: Dope too cold or viscous. The aircraft, dope and thinner should all be at the same temperature about 70°F, before spraying.

FABRIC WILL NOT TAUTEN: Fabric put on too loose. Fabric allowed to remain undoped too long. Too much retarder used for thinning dope.

BLUSHING: Humidity too high. Moisture in spray system. Dope applied over a moist surface.

PINHOLING OR BLISTERS: Water or oil in spray system or on surface. Undercoat not thoroughly dry. Too fast surface drying. Film coat too heavy.

BUBBLES AND BRIDGING: Dope too cold, or not brushed out properly. Temperature of dope room too high.

RUNS AND SAGS: Use of improper equipment. Improper adjustment of equipment. Improper thinning or faulty spray technique.

DOPE WILL NOT DRY: Oil, grease, or wax on surface.

DULL SPOTS: Porous spot putty or undercoating, allowing dope or lacquer to sink in.

BLEEDING: Organic pigments or dyes used in the undercoats which are soluble in the topcoat solvents.

WHITE SPOTS: Water in spray system or on surface.

PAINT OR PRIMER PEELING: Wax from stripper or detergent from cleaning process may still be on surface.

BROWN SPOTS: Oil in spray system.

ORANGE PEEL: Spraying with too high pressure. Use of too fast drying thinner. Cold, damp draught over surface.

WRINKLING: Reaction between solvents and primer or undercoats.

OVERSPRAY: Wrong spray technique. Too fast drying thinner.

FISH EYES: Silicone, wax or polish contamination on surface. Oil from air compressor.

CHAPTER 6

Landing Gear

Over the years different layouts of landing gear have been used, each with its particular advantages for specific circumstances. Aeroplane and helicopter landing gears may be divided into land, ski or water reflecting the different operating environments. In addition landing gears may be fixed or retractable, have a nose or tail wheel and, in the case of the helicopter, have either skids or wheels. But in each case the gear is designed to absorb the shocks of landing on a variety of surfaces, some more severe than others.

In order to absorb both landing and taxying shocks a variety of devices are used that prevent the energy being transmitted to the structure. The simplest of these is the spring steel strut now used in both flat and tubular forms for both aeroplanes and helicopters. In such a strut the tapering wall thickness and section allow the strut to act as a variable rate spring providing a progressive load deflection relationship. Many of the older light aircraft have used stranded rubber rope (known as a bungee) as the springing medium whilst yet others have used rubber blocks in compression.

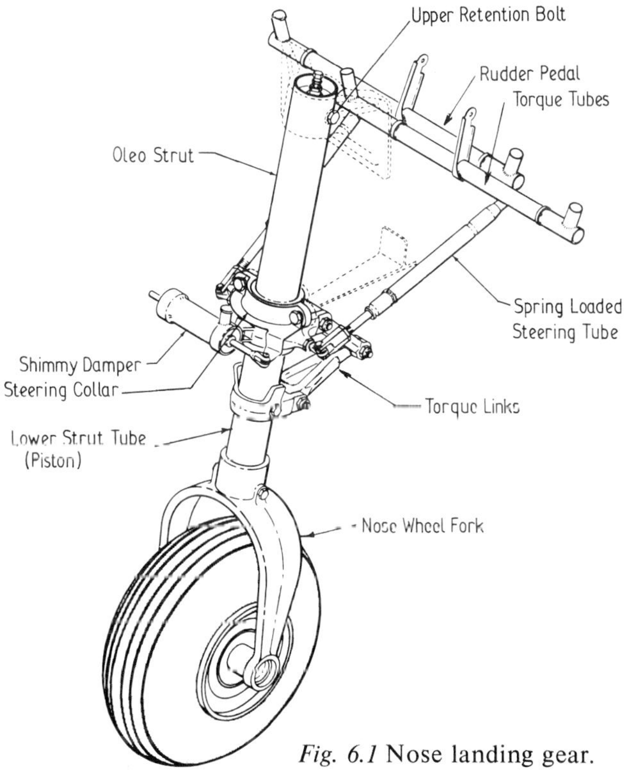

Fig. 6.1 Nose landing gear.

Probably the most common shock absorbing device is the oleo-pneumatic (air-oil) strut (fig.6.1) and the operation and servicing of this unit will be dealt with in some detail later. Another type of device has been used, known as the liquid spring, in which a piston is forced through oil under high pressure. However this is rarely met nowadays.

Fixed landing gear

The simplest form of landing gear is fixed directly to the aircraft structure. In some cases wheel spats or speed fairings are used to streamline the wheels in order to

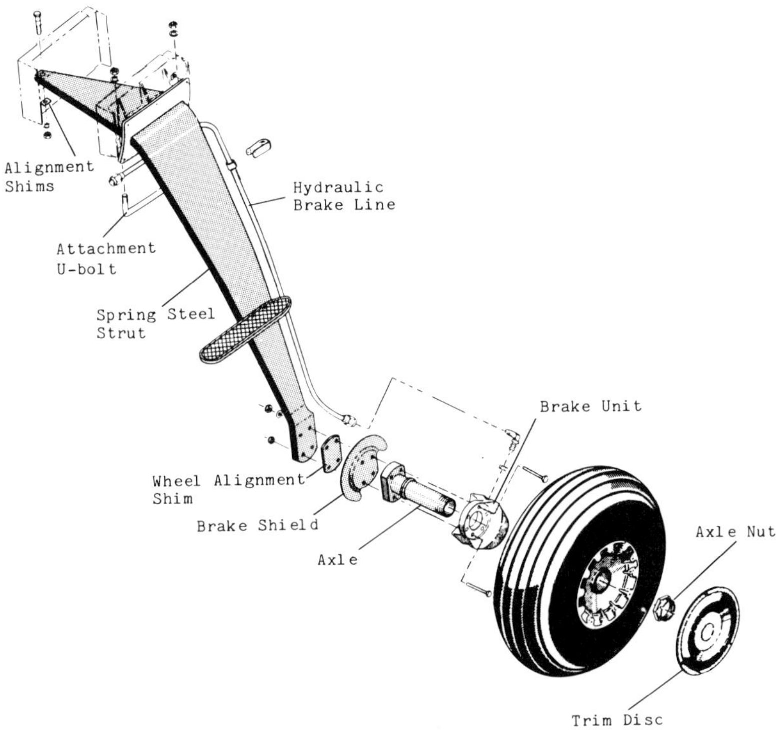

Fig. 6.2 Flat spring strut.

reduce drag. Flat spring struts (fig.6.2) are usually bolted and clamped to the structure, making a rigid assembly. Tubular spring struts (fig.6.3) are often mounted via a bonded rubber bushing which provides a better level of shock absorption. Tubular struts, or cross tubes, on helicopters are usually retained in some form of saddle with a bolted-on cap. Some helicopter skid gears are pivoted, however, and in this case the shock absorption is by means of dampers attached to the structure.

Periodic inspection of fixed landing gears must obviously include a check for security of attachment to the airframe structure. If loose bolts are found at any time they should be scrapped and replaced with new ones for there is every chance that they have been overstressed at some time. In view of the high stresses involved, along with the transmitted vibration, it is very important that only the correct torques are used when installing these bolts. From time to time the landing gear should be removed, stripped of paint and checked for cracks by any suitable method such as dye penetrant. Pay particular attention to areas where clips, brackets and steps are

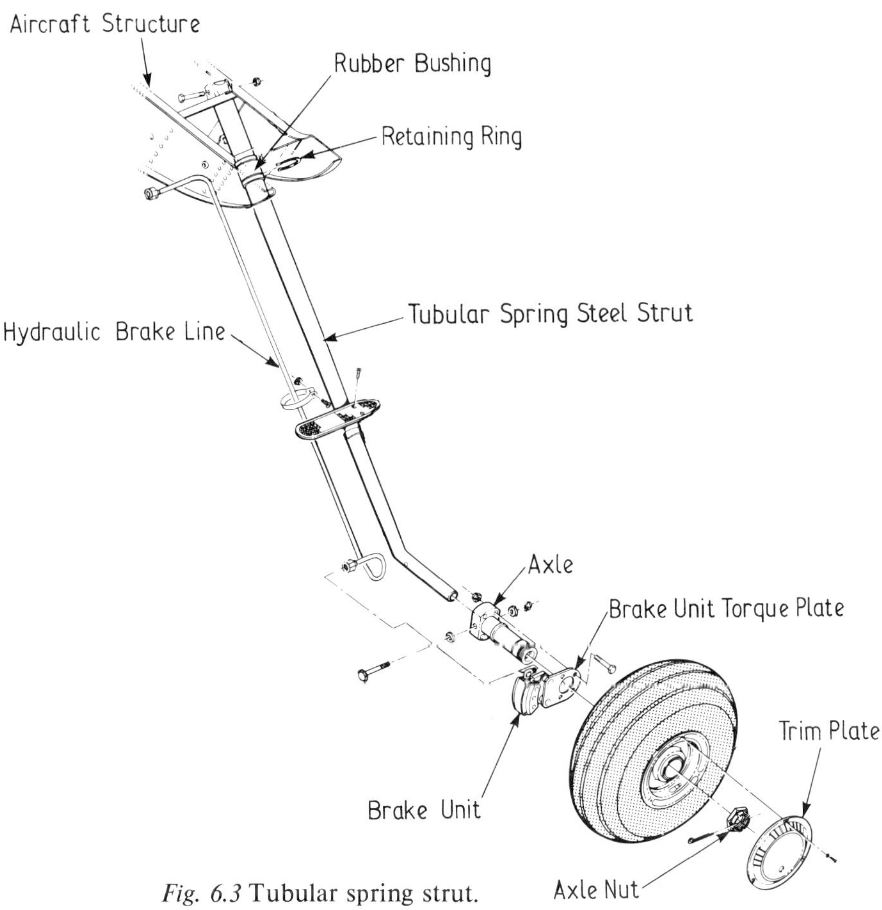

Fig. 6.3 Tubular spring strut.

attached or clamped. It is possible for moisture to creep under these items and cause corrosion. The corrosion pits provide points at which the reversing stress is concentrated and the result can be a sudden, brittle fatigue failure. Wherever possible such items should be sealed using an RTV type sealant (Room Temperature Vulcanisation), but only when it has been ensured that there is no entrapped moisture. Helicopter skids employ replaceable shoes to take the wear and these should be renewed when necessary (fig.6.4).

Float landing gear does not usually embody shock absorbers and is rigidly attached to the fuselage structure via a series of struts braced with streamline wires (fig.6.5). In most cases considerable reinforcement of the aircraft structure is necessary to take the high loads imposed. Inspection of float structure is much the same as any other structure except that it must be much more rigorous and searching. Damage from underwater obstructions or during beaching is a constant threat and thorough daily inspection is essential. The floats must be washed with fresh water daily and the bilges pumped out and inspected. Any bilge compartment found with a significant amount of water in it must be checked for leaks and repaired as necessary. Since corrosion is a constant threat all bilges must be treated with an inhibitor, for example sodium dichromate crystals in cotton bags. Any damage to paintwork must be made good at once and all fittings should be treated with a lanolin-based compound to prevent ingress of moisture.

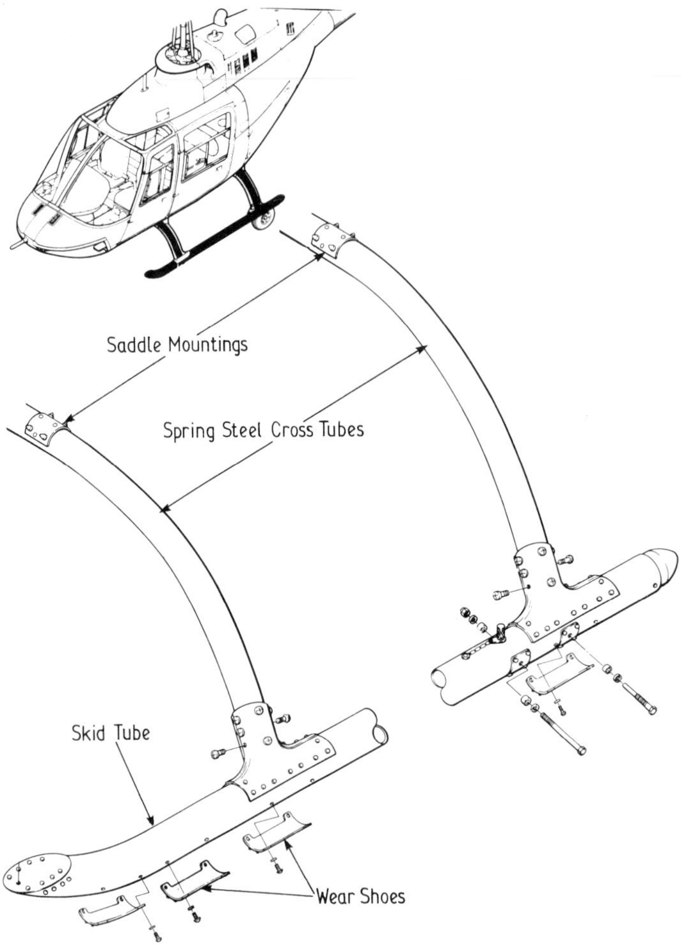

Fig. 6.4 Helicopter skid gear.

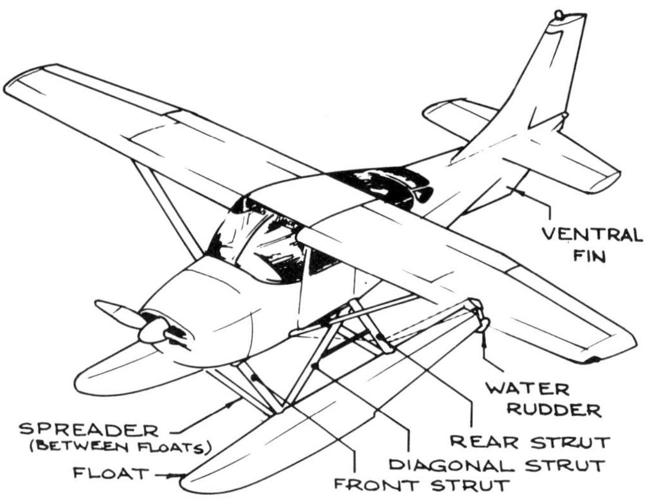

Fig.6.5 Float landing gear.

Removal and installation of fixed landing gear is straightforward and simple. Once the aircraft has been jacked and secured it is necessary to disconnect the hydraulic lines to the wheel brakes, capping them to exclude any foreign matter. In the case of nose or tail landing gear it will be necessary to disconnect the steering links or cables. Removal of the bolts will now free the landing gear for removal from the aircraft. Some flat spring landing gears are located by wedges at the outboard end and these must be driven out before removing the inboard bolts. This type of spring leg may incorporate washers under the inboard end which are used to level the wings when the aircraft is on the ground, and these must be counted and retained with the associated leg. Main landing gears utilising bungees as the springing medium fall into two categories in that some mount the bungees on a separate unit, often incorporating a hydraulic damper, whilst others connect the bungees direct to the leg

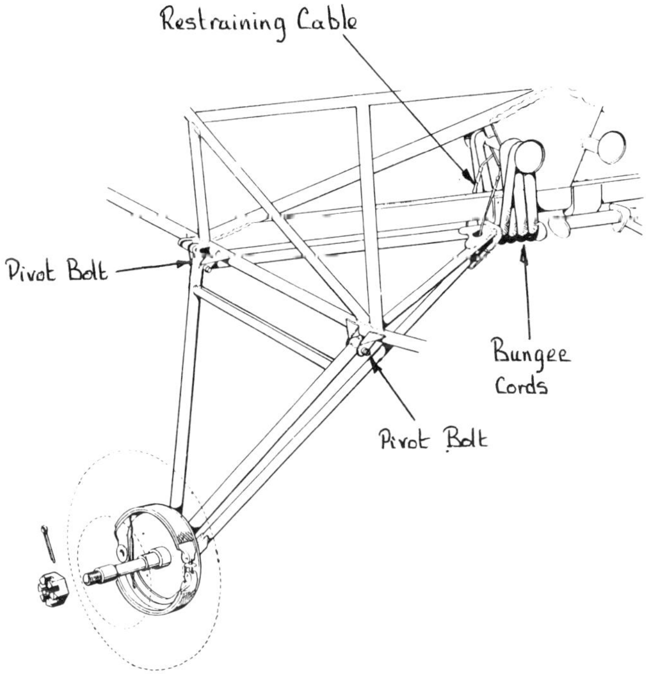

Fig. 6.6 Bungee installation.

(fig.6.6). In the former case removal of the leg is only a matter of removing the bolts but in the latter it will be necessary to remove the actual bungees using a special tool or a small crowbar which has been covered with PVC tape. A word of warning. Some nose oleo struts use the attaching bolts to retain the end plugs of the strut, so do always deflate the strut before removing it from the aircraft.

Installation is a reversal of the operations required for removal. A good policy is to renew all attachment bolts. The correct tightening torques must always be used. With many main landing gear installations it will be necessary to shim or adjust for a wings level position with the aircraft on the ground. With some landing gears it will be necessary to adjust the main wheel alignment, and with all it should be checked following installation. The procedure for carrying out this check will be discussed later.

Retractable landing gear

In order to reduce drag in the cruise, and thus improve cruising speed and fuel consumption, most high performance aircraft incorporate a retractable landing gear. A variety of methods of powering such a system have been used with the passage of time but the two most common utilise either electrical or hydraulic power. Where an electric motor is used to provide the motive power it usually drives a single actuator that operates all three landing gear struts (fig.6.7). A hydraulic system commonly employs a separate actuator for each strut.

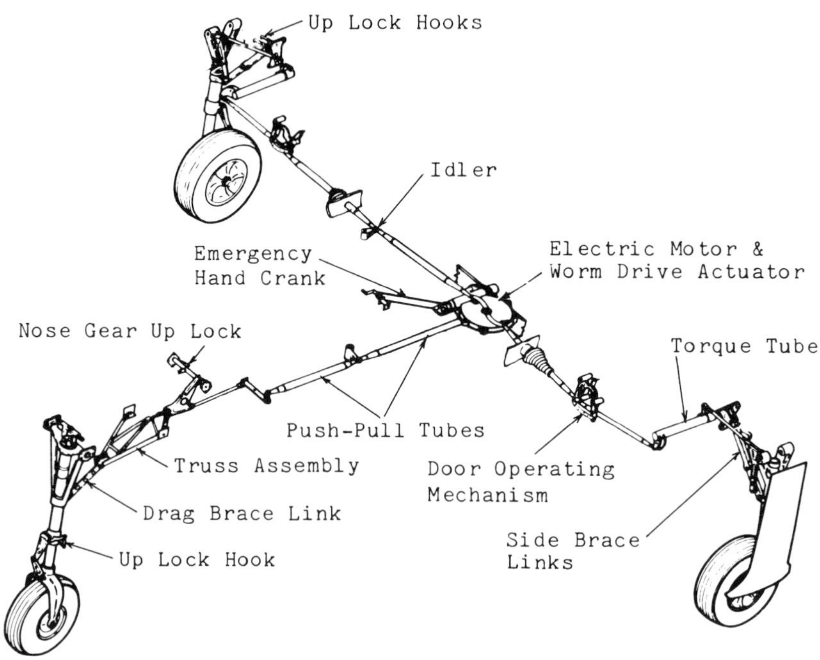

Fig. 6.7 Electromechanical retraction system.

Whilst there may be considerable differences in construction between one manufacturer and another most light twins employ a drag brace, functioning on the 'over-centre' principle, to maintain the strut in the down position (figs.6.8 and 6.9). By allowing a hinged strut to travel slightly past its centre position and contact an integral stop a mechanical lock is formed that is independent of the actual retraction mechanism. Thus loads transmitted up the landing gear strut are passed straight to the aircraft structure and not to the operating mechanism. Furthermore the landing gear will remain locked down in the event of a total power failure. An electro-mechanical system will also incorporate some manual means of lowering the landing gear in the event of power system failures. The majority of overcentre locks cannot be adjusted but it will still be necessary to measure the amount of overcentre to ensure that the lock is functioning correctly and that wear is not excessive. To do this a steel rule, or straight edge, is placed between the centres of the end pivot points. The distance of the centre pivot point from the steel rule may now be measured and compared with the dimension to be found in the maintenance manual (fig.6.10).

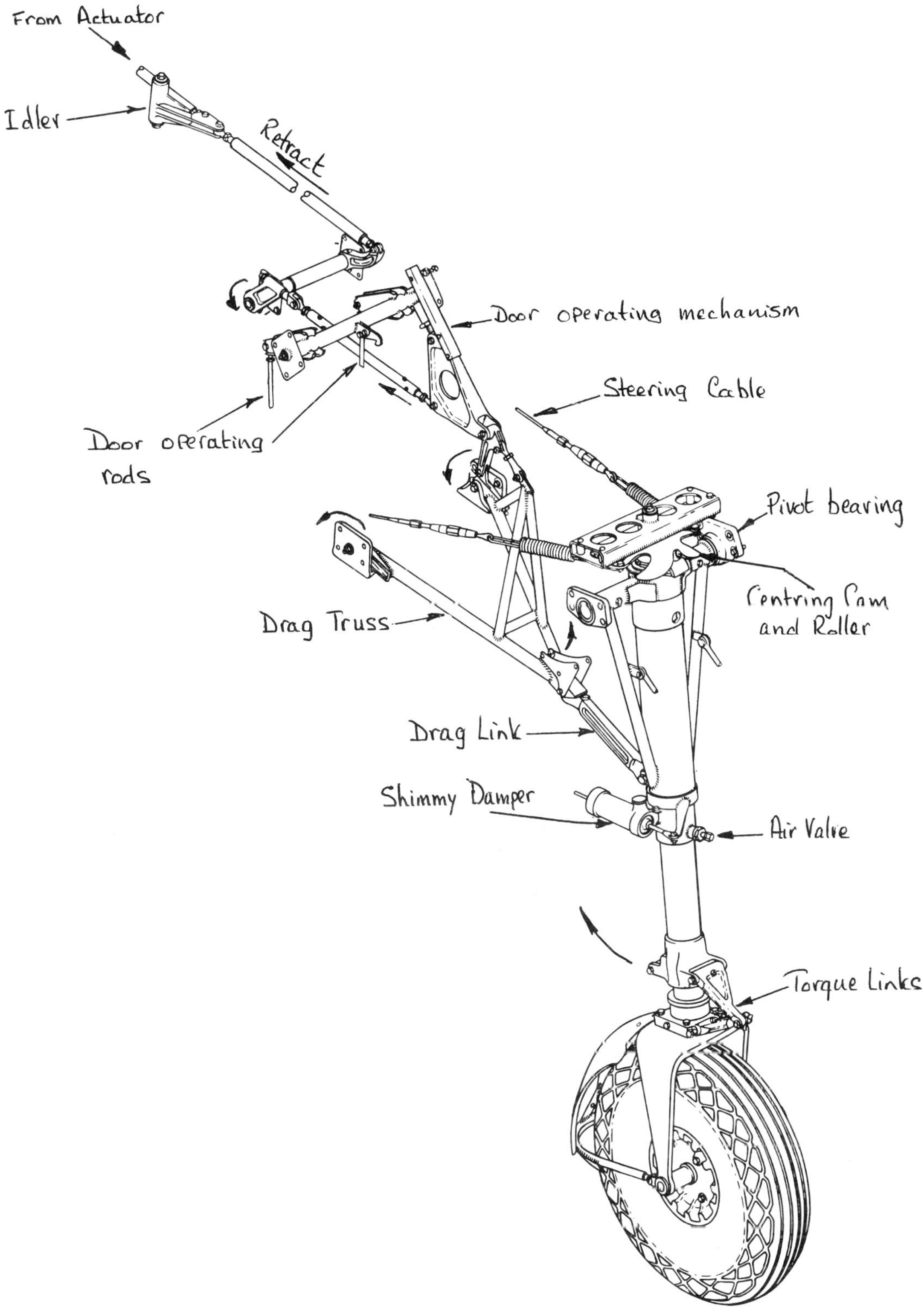

Fig. 6.8 Retractable nose landing gear.

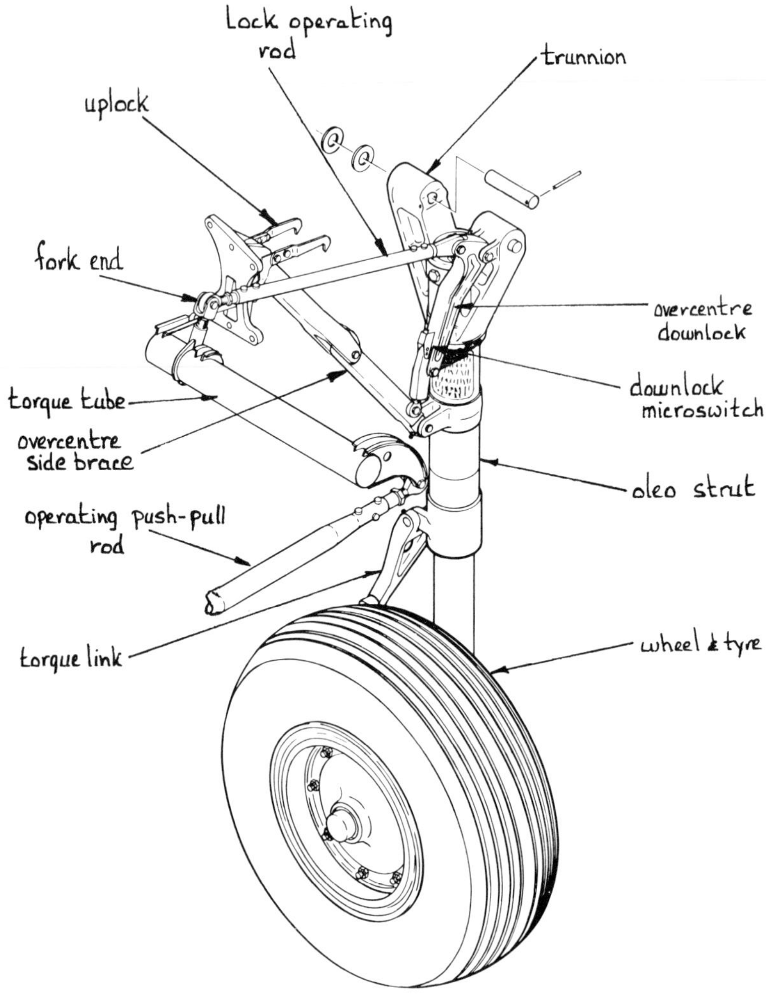

Fig. 6.9 Retractable main landing gear.

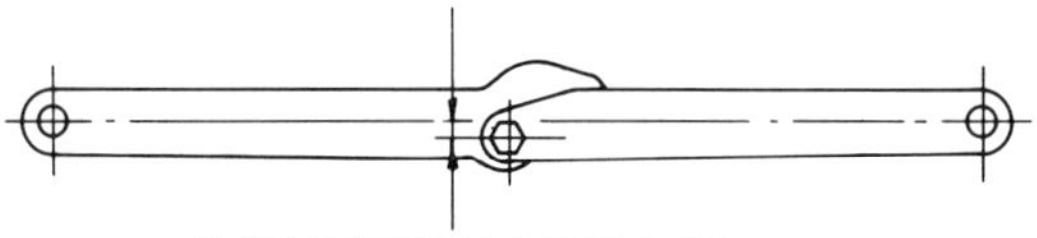

Fig. 6.10 Overcentre link.

Removal of retractable landing gear from the aircraft follows a procedure applicable to the specific type but there are certain principles to bear in mind. The first stage in any removal operation is to partially retract the landing gear in order to break the overcentre lock and to relieve the tension of any springs that may be fitted. It will then be necessary to support the landing gear in this position with a wooden box or something similar. In this way all loads are taken off the system and the bolts may be removed.

As the landing gear is dismantled a careful note should be made of worn parts and parts which have not been properly lubricated. Pivot bolts that show a dry coating of dirty grease may be caused by neglect or the grease nipple itself may be faulty. Make a note so that it can be rectified later. The landing gear mechanism may now be thoroughly cleaned to remove all accumulations of old grease and dirt. Drag braces should be paint stripped and checked for cracks, preferably using the dye penetrant method. Pivot points are provided with bronze bushes and these should be checked for wear against the dimensions given in the maintenance manual. Where no dimensions are given they should be checked for ovality and for clearance with a new bolt. Clearances should be of the order of 0.0035 in. in order to provide a passage for grease without at the same time creating excessive free play. Worn bushes may be replaced but may have to be reamed to size. Many bolts used in the landing gear and its attachments will be close tolerance and must never be replaced with any other type of bolt. Only ever fit bolts of the part number given in the parts catalogue and never guess at a suitable replacement.

Reassembly is the reverse of dismantling but again it is a good policy to use new bolts at all critical points. Check drag brace overcentre measurements before the landing gear is assembled in the airframe and use the correct torque when tightening all nuts and bolts. If grease nipples are replaced make sure that they point in a direction accessible to the grease gun when the landing gear is installed in the aircraft. Use the correct grease at all points and ensure that grease issues from both sides of the bearing when forced in under pressure. If grease issues from only one side find out why before proceeding any further. With the landing gear assembled to the airframe but before it is connected to the retraction mechanism swing the gear by hand and ensure that the overcentre lock engages correctly and that the strut is correctly aligned in the wheel well. The method of rigging the retraction mechanism is peculiar to each individual type but must be followed exactly *from the beginning.* Under no circumstances must a short cut be taken, particularly on electro-mechanical systems. To do so can result in a correctly operating mechanism but with high operating loads imposed on the components which will lead to early failure.

Once the landing gear is correctly assembled and rigged a full retraction test should be carried out. With a hydraulically powered system a ground power unit will be necessary and its characteristics and use are considered in Chapter 9. For an electrically powered system an external battery should be used in order to avoid discharging the aircraft battery. A further point to bear in mind is that if the voltage is allowed to fall the gear will not operate properly and a failure to recognise the problem could result in it being incorrectly rigged. The aircraft must be jacked sufficiently high to permit the landing gear to clear the ground during the retraction cycle, taking note of the fact that in some single-engined aircraft the struts actually descend before they start to retract. The purpose of a retraction test is to ensure that all parts of the landing gear function as they were designed, in the correct sequence and at the correct speed. The first retraction should be made in small stages or, if possible, at reduced speed so that any problems, such as lack of clearance between a wheel and its doors, may be seen before it causes damage. Only when all is seen to be in order should the operation be conducted at normal speed and the operating times checked. Whilst visually checking the operation of the landing gear listen for any unusual noises that may indicate incorrect rigging, fouling of parts or overloaded actuators.

Routine servicing of a retractable landing gear involves the usual inspections for security, condition and correct adjustment. However there are two other items that are vital for correct, consistent and reliable operation. One is lubrication and the other is cleanliness. Neither is difficult to achieve satisfactorily but many failures of retractable landing gears have been attributed to deficiencies in one or the other areas. Both are of paramount importance with amphibious landing gears. Lubrication has already been mentioned in the context of assembly but in service a further point arises. When lubricating the landing gear mechanism retract the gear sufficiently to break the locks and thus relieve the loads on the pivot bolts. Then grease in the normal way until fresh grease issues from both sides of the bearing. Cleanliness is important in that any accumulations of dirt will collect moisture that will promote corrosion and possibly cause electrical microswitches to malfunction. Overcentre locks will not operate if there is dirt on the mating faces and the gear will not lock down. Clean the drag brace with the gear partially retracted so that you are sure that the mating faces are really clean. A final word of caution. If an oleo strut is not correctly inflated *do not retract the gear.* In many installations the wheel will foul the doors or jam the uplock, causing a major mechanical failure that the emergency lowering system will not cope with.

So far nothing has been said about the operation of wheel well doors. In some aircraft these operate mechanically, the simplest method being an adjustable link to the landing gear strut or to another part of the retraction mechanism. Such a door usually stays open whilst the landing gear is down, although a refinement of the mechanism allows the door to close once the gear is down. In hydraulically operated systems separate jacks may be used to close the doors and the operation of these is discussed in Chapter 9. In all cases doors must be adjusted so that they fit snugly when closed in order to avoid any disturbance of the airflow. Free play in the hinges and mechanism must be kept to a minimum in order to avoid problems caused by vibration.

The oleo strut

The oleo strut, or air/oil strut as it is sometimes known, is probably the most popular form of landing gear strut in use. The air, being compressible, provides the springing medium as the two halves of the strut telescope over each other. As the strut shortens under load so the pressure of the air rises and increasingly resists the load applied. Removal of the load under these circumstances would result in a nearly instantaneous extension of the leg to its full length giving a very bouncy ride. The purpose of the oil is to damp the movement of the strut and to do so progressively. The nature of the damping is such that there is a greater degree of damping at the end of the strut travel than when the strut is in an intermediate position. Consideration of a simple strut will show this more clearly.

As the strut compresses the oil is forced from the lower chamber to the upper chamber through the orifice, the size of the orifice being determined by the position of the tapered metering pin (fig.6.11). As the compression continues so the orifice is reduced in area and the damping increases. In extension the reverse is true until the end of the stroke is neared. At this point the increasing diameter of the metering pin again reduces the area of the orifice and increases the damping. In such a strut the air and oil are free to mix but a variation in design found in some of the larger twin-engined aircraft utilises a floating piston to separate the air from the oil. The oil is still metered although in this case a multi-orifice snubber is used. The operation of a snubber is illustrated in figure 6.12, although the strut shown does not have a separator piston. On compression of the strut the oil flows through all the holes in the

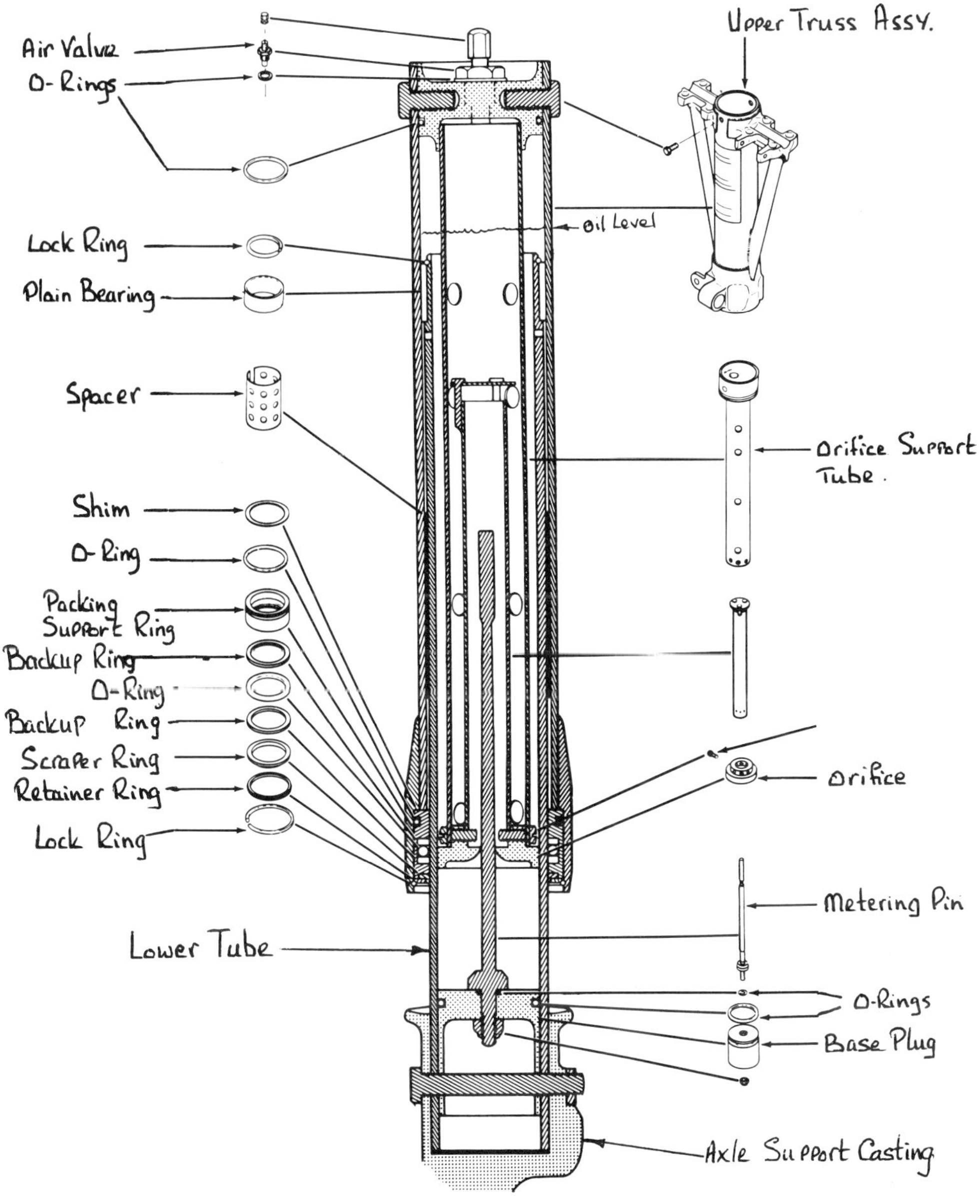

Fig. 6.11 Oleo strut – with metering pin.

snubber and forces the separating piston down, compressing the nitrogen charge. On the return stroke a plate valve is closed by the oil flow, assisted by a spring, and closes all the snubber orifices with the exception of one. Thus the oil flow is restricted and the speed of the strut extension is reduced. The snubber valve is also used on oleo struts without the separating piston. Torque links are used on all oleo struts to prevent rotation of the tubes and so maintain wheel alignment. They also provide a

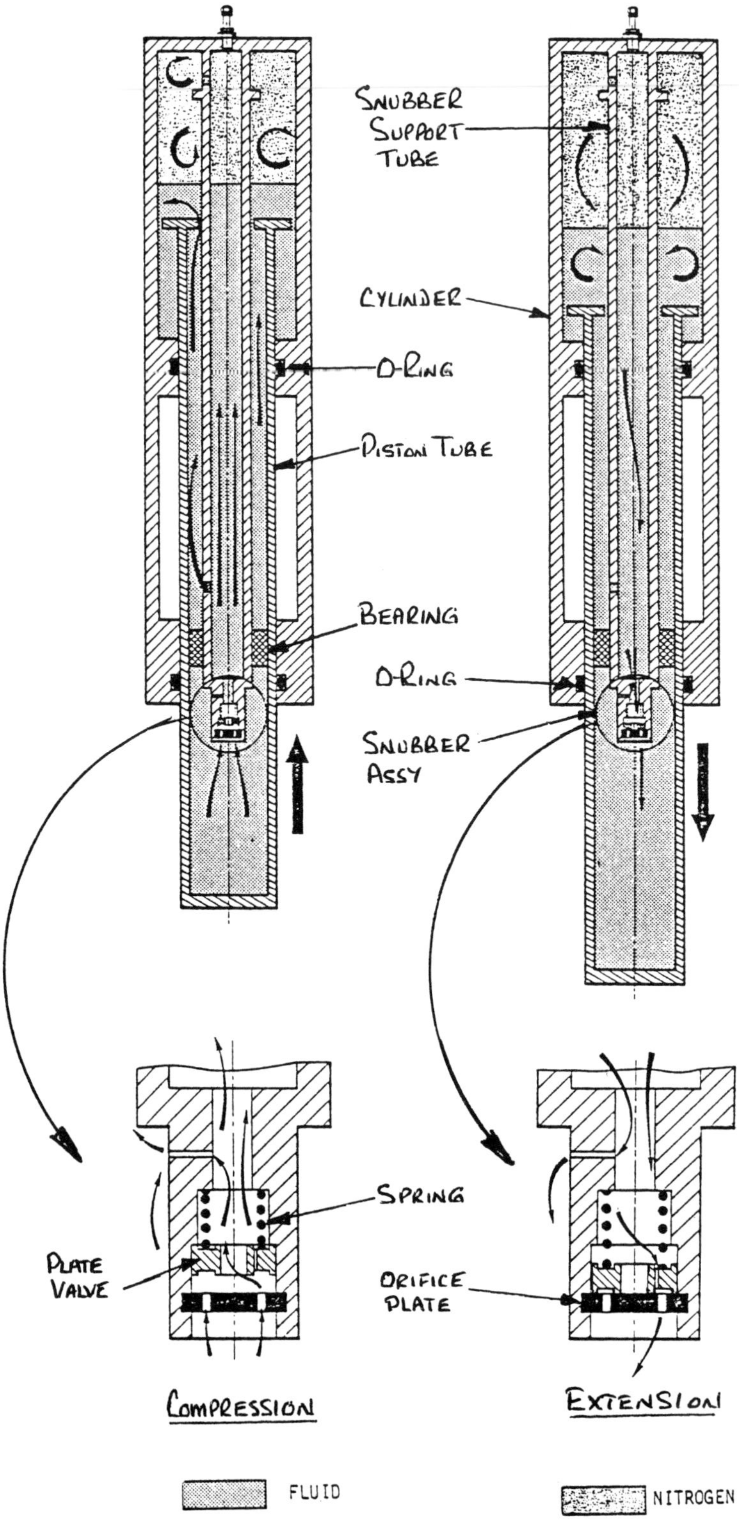

Fig. 6.12 Oleo strut – with snubber.

stop to limit the extension of the strut and, in the case of the nose strut of many light single-engined aircraft, the upper link is used to locate the nosewheel in its central position with the strut extended.

The procedure for dismantling a strut varies from one type to another but there are certain precautions and procedures they all have in common. Cleanliness is number one precaution and whether the strut is on or off the aircraft always clean it thoroughly before dismantling. Dirt and hydraulic components do not mix and the most common cause of failure is dirt working past the scraper ring and damaging the O-ring seal. Number two precaution is to deflate the strut, by gently depressing the valve core, before doing anything else and then remove the valve core. If a strut is being dismantled on the aircraft it is usually necessary to remove the wheel in order to obtain sufficient clearance to separate the two halves of the strut. Do arrange a clean drip tray to catch the oil from the strut as it is dismantled and do have a clean surface on which to place parts that are ready for assembly.

Once the strut is dismantled remove and discard all O-rings, seals and scraper rings. Take the opportunity of inspecting the component parts of the strut for wear and damage paying particular attention to the orifice, metering pin, orifice support tube and plain bearings. If a steering bearing is involved, as on a nose strut, this is the time to check it for proper adjustment and lubrication. Lubricate all parts and seals in the correct clean hydraulic fluid before reassembling. Be very careful when reassembling to avoid damage to the seals and scraper rings; use plenty of hydraulic oil as a lubricant. Once assembled check that the strut operates freely without either binding or excessive play. Filling the strut varies with the design but in general the strut is compressed to within 2 in. of the end of its travel. Whilst it is held at this point it is filled with hydraulic oil through the valve at the top (with the core still removed). A clear PVC tube is attached to the valve and immersed in a container of clean hydraulic fluid and the lower strut is moved up and down over a travel of 6-8 in. finally ending with the strut fully compressed. The PVC tube is removed, the strut extended and the valve core refitted. It may now be inflated to the required air pressure. A foot pump will be sufficient for the nose strut of many light aircraft but a compressor or nitrogen cylinder will be necessary for some others. When using a nitrogen cylinder attach the hose to the strut and make sure that the regulator screw is backed out as far as it will go. *Very slowly* open the valve on the cylinder and then screw in the regulator until the required pressure is obtained. Close the gas cylinder valve and remove the hose from the strut slowly in order to dissipate the gas pressure in the hose. Compressed air may be used as an alternative to nitrogen but nitrogen has the advantage of being dry.

In service the oleo strut requires little attention. Always keep the exposed portion of the lower strut clean; simply wipe clean with a cloth. Check that the extension is correct and never operate the aircraft with a deflated strut. Check the damping action by pushing down on the nose of the aircraft, releasing it smartly and making sure that it damps within about 1½ cycles. This is more difficult to check for the main landing gear but experience will teach what is right for a specific aircraft.

The tail wheel

There are varieties of tail wheel as there are varieties of any landing gear. However most are steerable and incorporate a shimmy damper. In the unit illustrated (fig. 6.13) the shimmy damper consists of two Tufnol washers of different thicknesses that

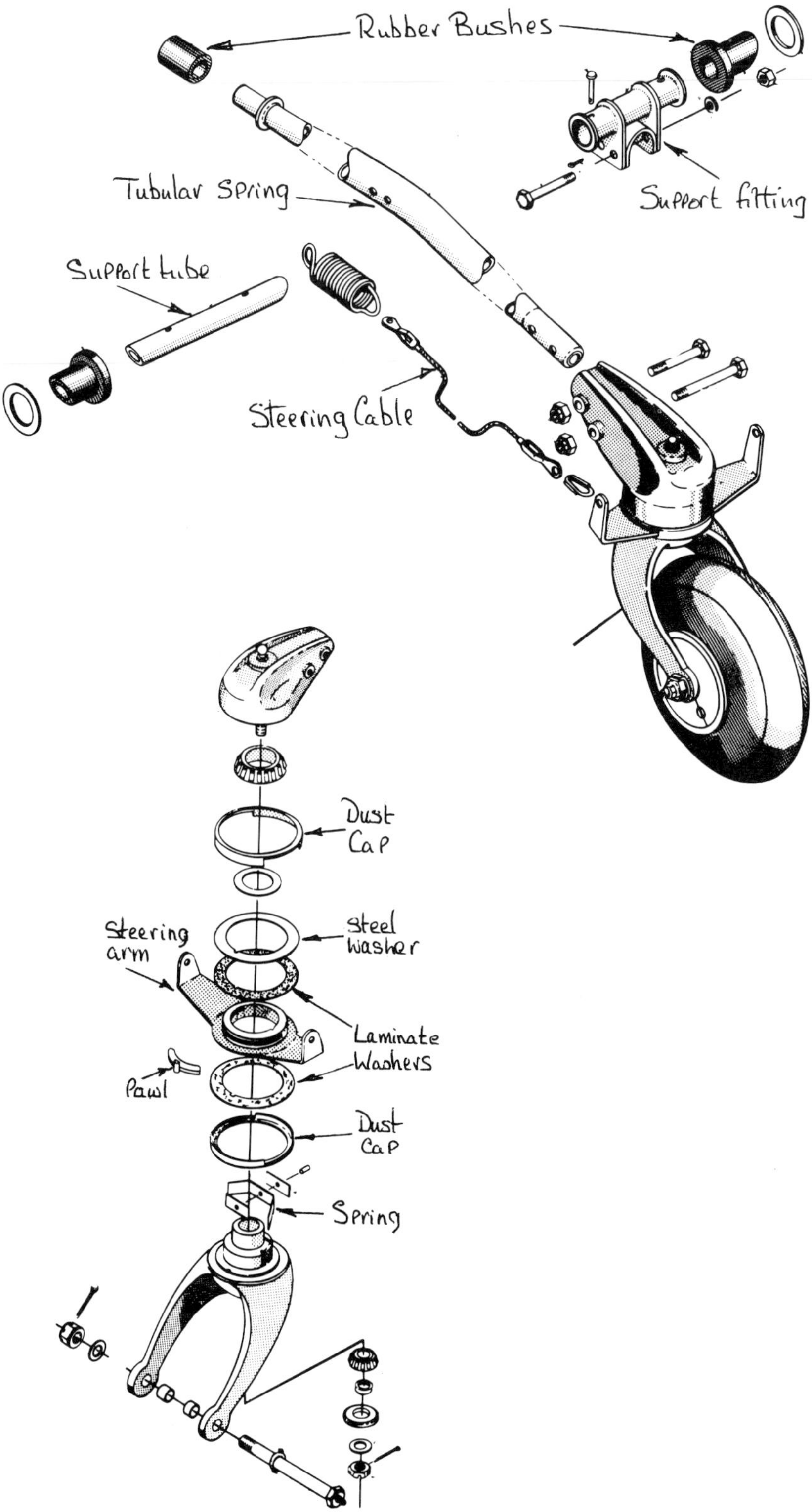

Fig. 6.13 Tail wheel installation.

are selectively assembled alternately with steel washers. The upper laminate and the steel washer are keyed to the tail wheel steering arm via a pawl whilst the lower laminate washer is keyed to the tail wheel fork. The washers are selected to maintain a clearance of 0.004–0.024 in. (0.1–0.6 mm) in the assembly. From 3 to 10 springs may then be fitted in the support bracket to provide the amount of damping action required.

Since tail wheels are usually fitted to 'bush' aircraft they tend to lead a rather hard life and thus should receive a greater amount of attention than larger wheels. The effectiveness of the shimmy damper can be checked by disconnecting the cables from the steering arm and conducting a 'pull off' check. In this a spring balance is attached in place of one of the cables and the force required to start the assembly moving is measured. The manufacturer gives a minimum value for this which would often be of the order of 5–6 lb (2.3–2.7 kg). Should it not be possible to achieve this value the unit must be dismantled, cleaned and inspected. The fibre washers wear and the keyway in them tends to tear with extended use. All worn parts should be renewed, the bearings lubricated and the clearance checked.

The tail wheel is steered by cables that are connected to the rudder system. Thus as the rudder is moved so the tail wheel will turn in the appropriate direction. It is usual to incorporate springs in these cables to prevent the rudder system being overloaded at times when the tail wheel is not free to swivel. The ability to steer is limited to about 24° either side of centre. On reaching this limit the pawl is disengaged from the spring and the fork becomes free swivelling. Usually adjustment of the tail wheel steering is not necessary due to the springs incorporated in the system. However during the course of routine maintenance, when the tail of the aircraft is clear of the ground, the tail wheel assembly should be checked for correct alignment with the rudder in the neutral position.

Nose wheel steering

Nose wheel steering, in the typical light aircraft, is also connected to the rudder system but direct to the rudder pedal assembly. The steering arm on the nose strut is connected to the rudder pedal torque tubes by two telescopic spring loaded struts (fig.6.1). When the nose oleo strut is fully extended a block on the upper torque link locks the nose wheel in its neutral position, but the spring struts permit the normal operation of the rudder and provide it with a spring-loaded neutral position. In many aircraft it is possible to offset this neutral position and so provide a small degree of rudder trim. In the retractable gear aircraft the same principles still apply but there is a more complicated device for centring the nose wheel when the gear retracts. This often consists of a roller and cam arrangement located at the top of the oleo strut which is connected by cables and springs to the rudder pedals (fig.6.8).

Rigging of the nose wheel steering system must be carried out in conjunction with rigging of the rudder control system. It is essential that the rudder and the nose wheel are in line in the neutral position. In the typical light single-engined aircraft the nose wheel is rigged first and then the rudder. However in many of the twin-engined aircraft the reverse is true because the nose wheel system can be adjusted independently.

Hydraulic shimmy dampers

The majority of nose landing gears that incorporate an oleo strut also utilise a

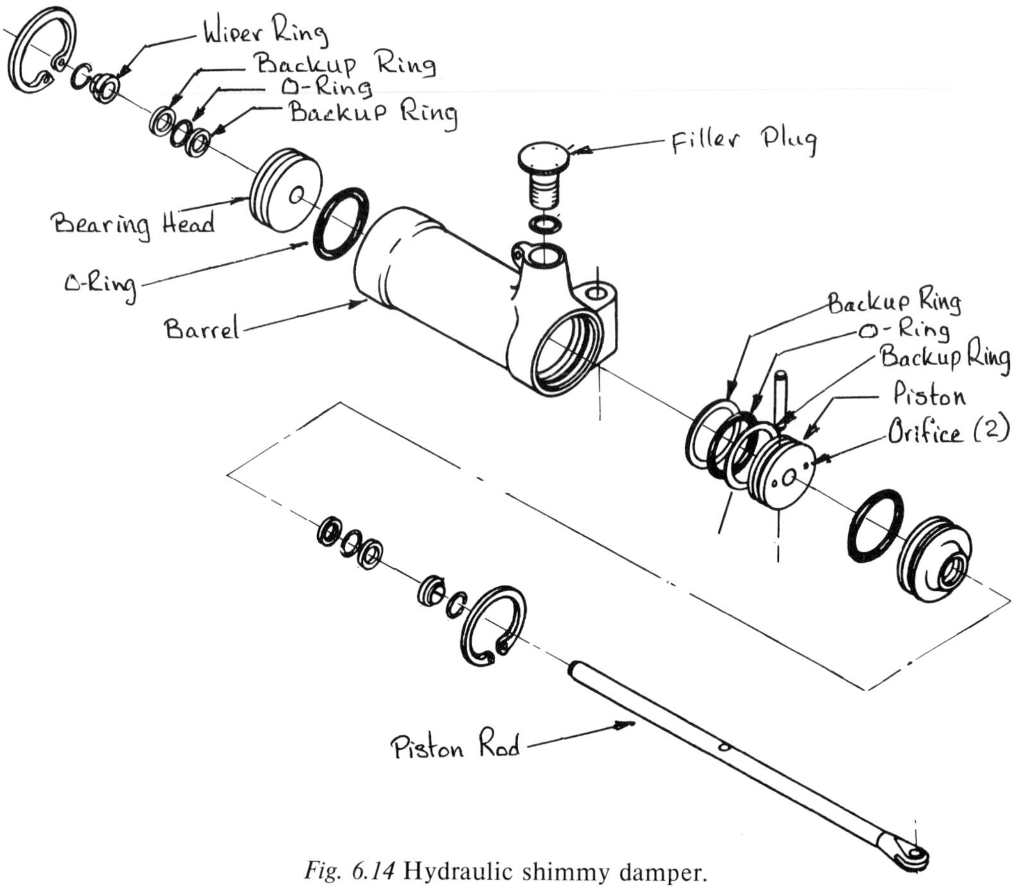

Fig. 6.14 Hydraulic shimmy damper.

hydraulic shimmy damper (fig.6.14). The barrel of the shimmy damper is filled with hydraulic fluid and movement of the nose wheel forces the piston rod and piston through the fluid, which flows through orifices in the piston. The efficiency of the shimmy damper may be checked by a simple practical test. First place grease plates under the nose wheel and then attach the steering arm used for normal ground handling. Movement of the steering arm to the right and the left should be met with a noticeable amount of resistance. Should this not be the case the damper should be removed and the cause ascertained. Usually fitting new O-rings and servicing with fresh hydraulic oil is all that is required. After repair the damper must be filled with hydraulic oil and completely bled of air in the manner recommended by the aircraft manufacturer. To achieve this some dampers have to be exercised whilst completely submerged in oil but others only require filling in a vertical position.

Wheel alignment

If abnormal main wheel tyre wear occurs it could be that the main wheel alignment is incorrect. In any case it would be necessary to check wheel alignment after a heavy landing or after a major repair, even on an aircraft when there was no provision to adjust it. The procedure is simple and requires very little in the way of equipment. The aircraft should be ballasted to its normal operating weight and rolled forward so that the main wheels rest on two sets of grease plates (fig.6.15). A set of grease plates

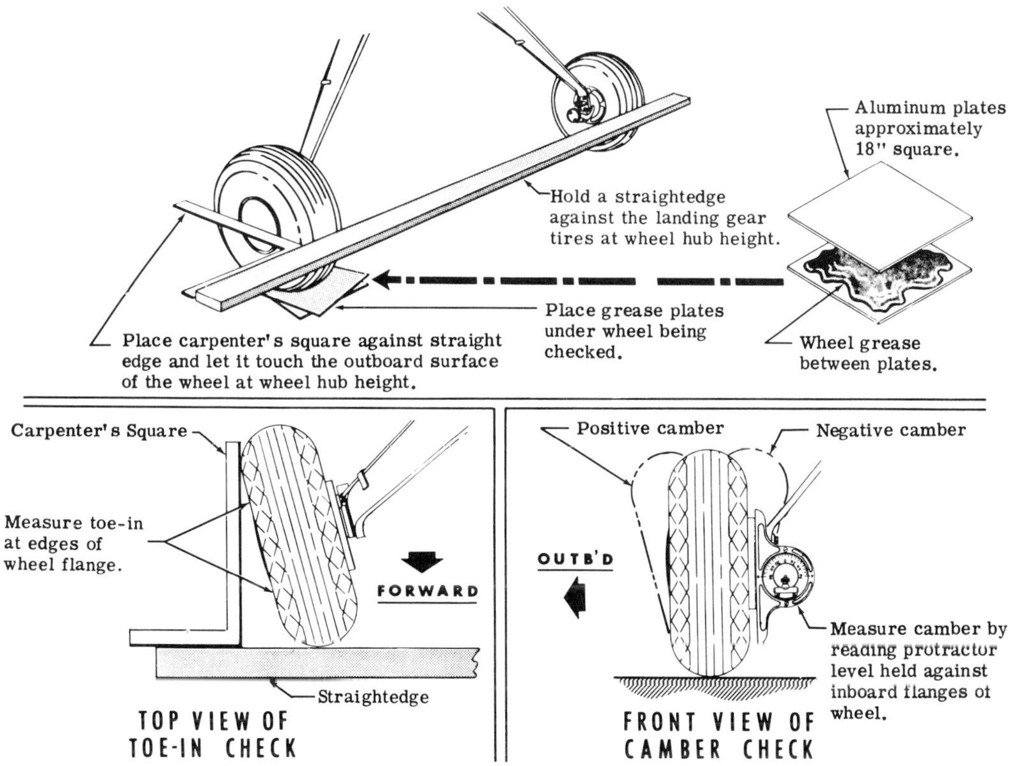

Fig. 6.15 Landing gear alignment check.

is simply made from two pieces of sheet metal about 18 in. square. Any suitable wheel-bearing grease is spread between the two plates, thus providing a mobile surface that will allow the landing gear to take up its natural position. A straightedge should now be placed in front of the wheels and supported so that it just touches the outside diameter of both tyres a little below axle height. A large carpenter's square, or a roofing square, is used to check the toe-in of the wheels by measuring the gap between the wheel flange and the square. The toe-in measured at one wheel should be half the total figure quoted in the aircraft specification. The camber may be read by placing a bubble protractor against the flange of the wheel as shown. The camber is said to be positive when the top of the wheel leans outwards and negative when it leans inwards. The landing gear may now be adjusted and rechecked or the cause of the misalignment traced.

Helicopter ground handling wheels

Helicopters that are fitted with skid-type landing gear are provided with some form of detachable wheels to enable the helicopter to be moved on the ground. These are simple industrial type wheels that are very similar to some tail wheels. No special maintenance techniques are involved.

CHAPTER 7

Wheels, Tyres and Brakes

The type and size of wheel and brake assembly is closely related to the weight of the aircraft and its landing speed. As the size of the aircraft increases so the energy dissipated in landing the aircraft and bringing it to a stop increases, resulting in increased size and complexity of the components. However within the class of aircraft covered by this book wheel and brake equipment falls into several clearly defined groups, each of which will be discussed in this chapter.

Types of wheels

Irrespective of make modern wheels fall into two distinct groups, namely divided wheels (fig.7.1) and loose flange wheels (fig.7.2). For many years the divided wheel

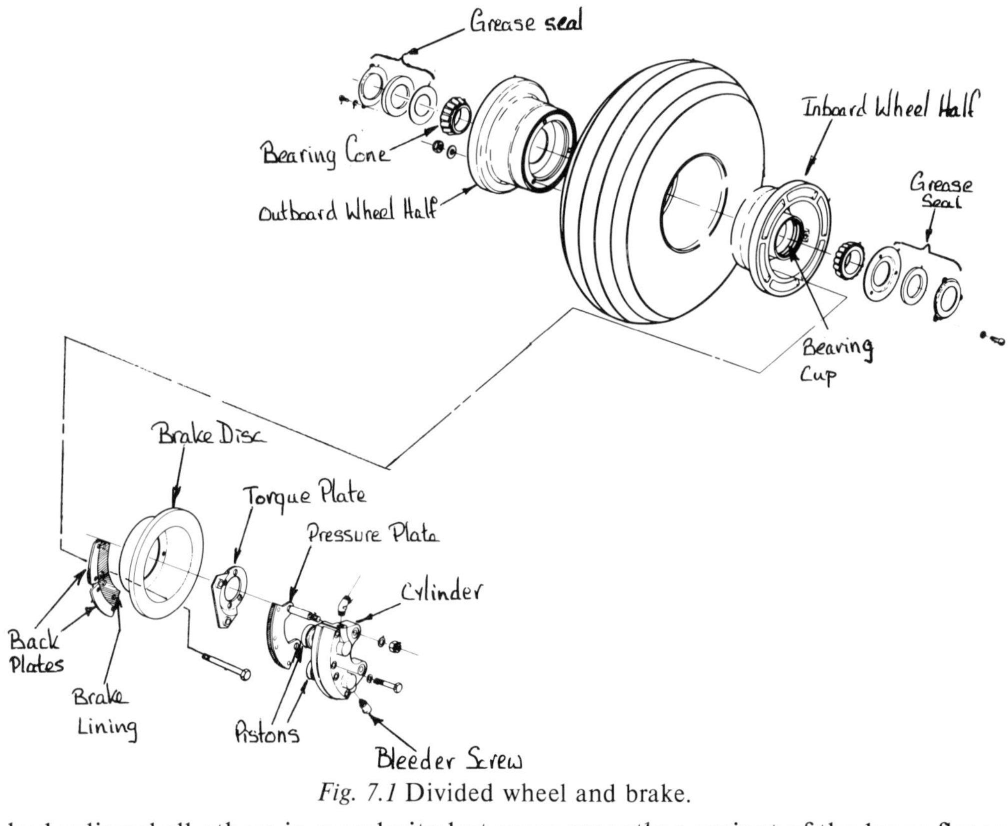

Fig. 7.1 Divided wheel and brake.

had eclipsed all others in popularity but more recently a variant of the loose flange wheel has gained some ground on light single-engine types. The divided wheel

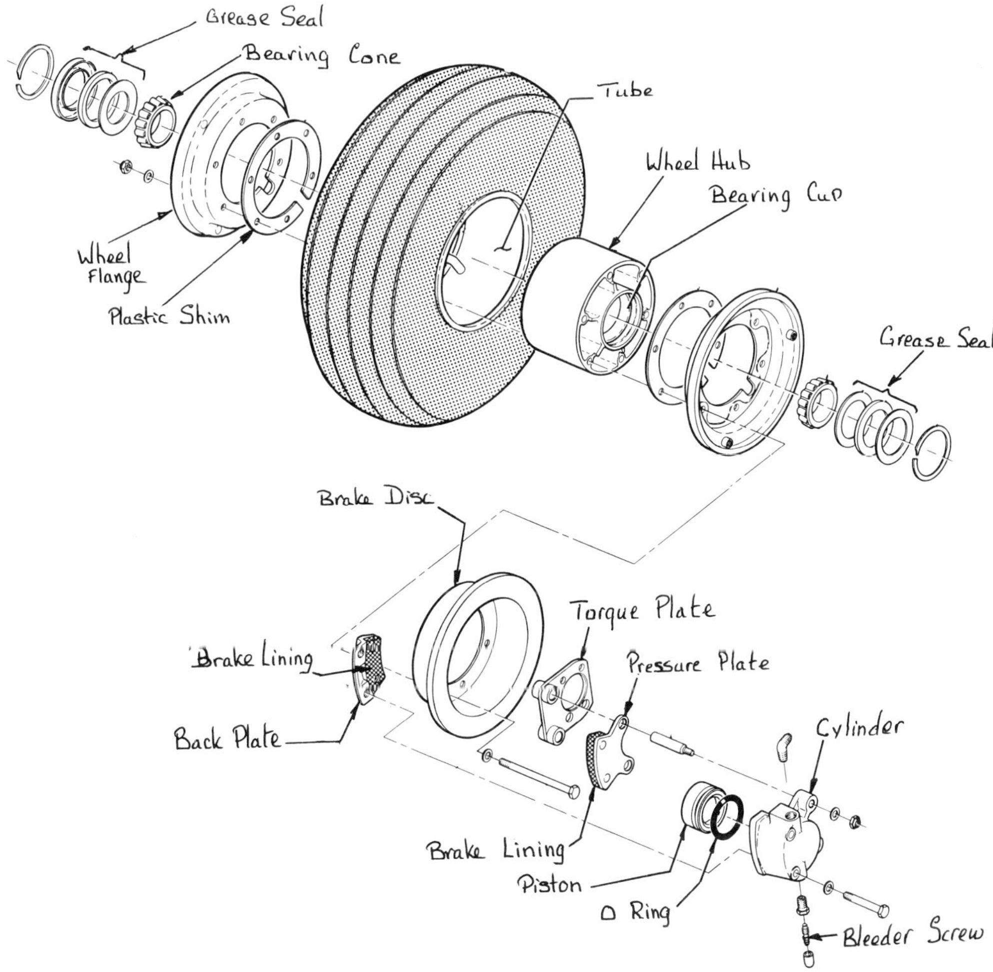

Fig. 7.2 Loose flange wheel.

consists of two halves that are held together by bolts passing through the web of the castings. The mating faces of the two halves are machined in such a way that one spigots inside the other in order to provide positive location and alignment. Traditionally the loose flange wheel has consisted of a hub cast integrally with one of the flanges. The other flange has then been machined to slide over the hub where it is retained in position by a lock ring (fig.7.3). The modern edition of the loose flange wheel consists of a cast light alloy hub to which are bolted two steel flanges. Plastic shims are used between the flanges and the hub to prevent galling and corrosion due to the proximity of two dissimilar metals. The main advantage of this type of wheel is that the tyre can be changed (in the case of a main wheel) without removing the brake or the wheel from the axle. However difficulty can be experienced in breaking the tyre beads with the wheel still on the axle unless a pincer type of bead breaker is used.

Wheel removal

Whilst there may be small variations from one aircraft to another the basic procedure for removal of a wheel is the same. First it will be necessary to jack the aircraft

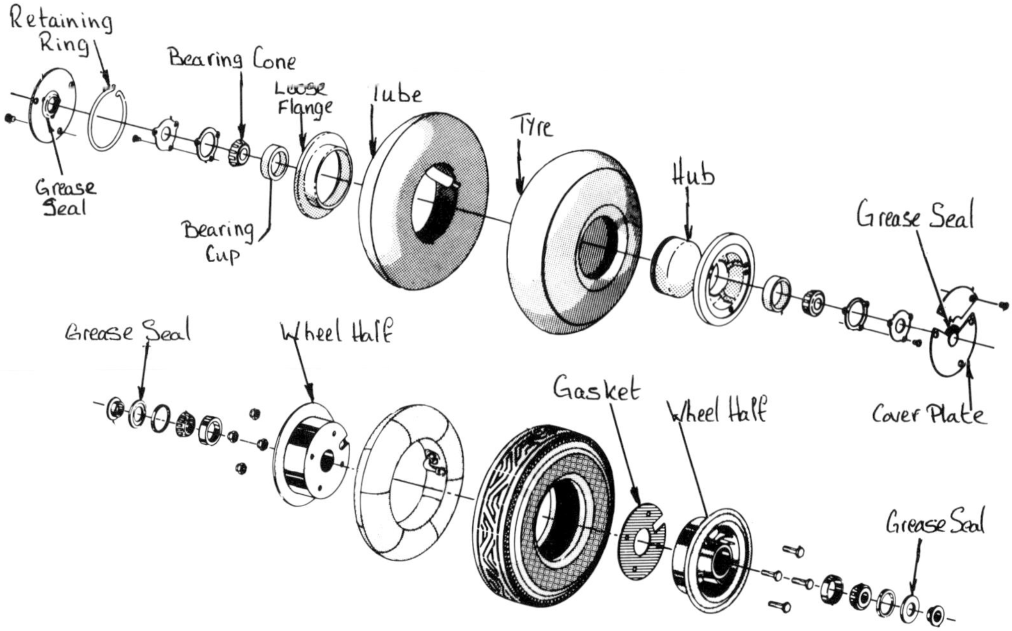

Fig. 7.3 Loose flange and divided tail wheels.

properly and the procedure for this is described in Chapter 1. The next step is to release the brake assembly where either an expanding shoe brake or a fixed type disc brake is employed. In the former the shoes should be adjusted away from the drum and in the latter it will be necessary to unbolt the back plates allowing the linings to drop clear. Where a loose disc type of brake is fitted it may be necessary to prise out the disc spring clips in order to free the disc. The wheel dust cover and the hub cap should now be removed and placed to one side in a tray which will subsequently contain all the parts removed from the one wheel. The split pin (cotter pin) should now be pulled from the axle nut after the legs have been straightened. Whilst there are special puller tools on the market a large pair of diagonal cutters may often be used to pull the pin with more effect. The axle nut may now be unscrewed and the wheel withdrawn from the axle. In some designs the wheel bearings will drop free as this is done, so take care to catch them. In other designs they are retained in the wheel hub by the grease and dust seal retainer.

Dismantling the wheel

As with any other maintenance operation there is a right and a wrong way to dismantle the wheel. The problem in this case is that if the wrong way is chosen it can result in injury to the mechanic. ALWAYS DEFLATE THE TYRE FIRST. If the tyre has been inflated to 30 p.s.i. then there will be a force of 30 lb for every sq. in. of bead area that is trying to push the two halves of the wheel apart. It could be dangerous so do take the necessary precautions. Remove the valve dust cap and the valve core using a proper tool. If the tyre is a high-pressure tyre the pressure should be bled off before removing the core. In the absence of a commercial bead breaking tool (fig.7.4) the wheel may be laid on a sheet of plywood, or on a bench, and the bead broken from the wheel using a tapered hardwood drift and a heavy hammer. With the tyre free from the wheel the through bolts may be removed, the wheel halves or

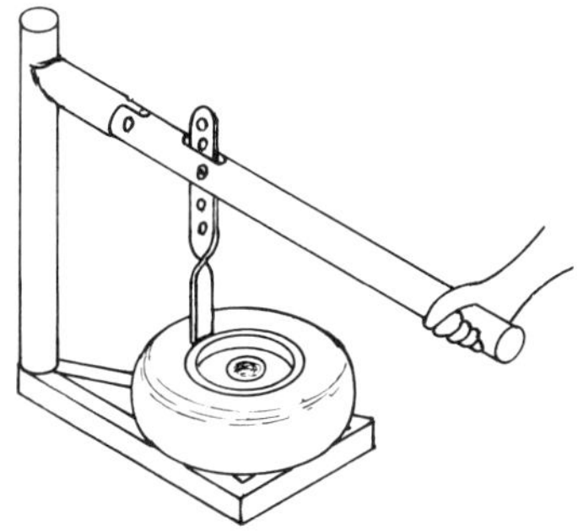

Fig. 7.4 Tyre bead breaker.

flanges separated and the tyre and tube removed. The bearings and seals may now be removed from the hub.

Wheel inspection

Before any other steps are taken a visual inspection should be made of all the parts and any obviously unserviceable items put to one side for renewal. The remaining parts should be cleaned and spread out on a clean bench for further inspection. The wheel halves, hubs and flanges should be visually inspected for signs of cracking, corrosion, distortion or other damage. Cracks are most likely in the vicinity of the through bolt holes and the beads but do not let this blind you to damage elsewhere. Minor damage to the rim may be sanded smooth provided that the repaired area does not include part of the rim where the tyre bead normally seats. Minor corrosion should be dealt with in a suitable manner and the repaired areas reprotected with a primer followed by silver lacquer. Where a wheel contains drive keys or teeth for a brake disc these must be carefully inspected for wear. The manufacturer's tolerances must not be exceeded or faulty brake operation will result. Worn and damaged drive keys may be replaced but if drive teeth are damaged or worn it will be necessary to replace the wheel half. The bearing outer race (cup) should be inspected for security in the hub and for signs of wear, scoring or overheating. Should it be necessary to replace the outer race the wheel hub is heated in hot water in order to expand it and the race replaced whilst the hub is still hot. The bearing inner race (cone) should receive a similar inspection after thorough cleaning in clean solvent. The inner race should be lubricated in the specified grease using a bearing greaser but if this is not available then it may be packed by hand. Whilst it is important that grease is forced between the rollers and the cage it should not be overpacked or the bearing will overheat and exude grease at a later stage in service. The cleaned and lubricated bearing should be kept in a plastic bag until it is required for use. Felt grease seals should be renewed.

Wheel assembly

Re-assembly of a wheel is very much the reverse of dismantling, but there are some very important points that need to be attended to during the procedure. Details of the tyre and tube will be explained in a later section. The inside of the tyre will be dusted with french chalk (to act as a lubricant for the tube) and the lightly inflated tube placed inside. It will be necessary to align the balance marks on tyre and tube at this stage. Place the outboard wheel half in the tyre taking care to position the valve stem through the valve hole. Insert the through bolts, and brake disc where fitted, into the inboard wheel half and position in the tyre to align with the outboard half already fitted. The washers and nuts may now be fitted and tightened using a torque spanner. In many cases the torque is cast on the wheel halves but any figure quoted in the Service Manual must be taken as the overriding value. The bolts should be tightened equally, a small amount at a time and diagonally, or in any order specified in the manual. Uneven or incorrect torque of the through bolt nuts may cause bolt failure. When bringing the wheel halves together be careful to avoid pinching the tube between them. There should be a definite metallic click as the two halves contact each other, showing that there is no interference from the tube.

Proper inflation is the most important maintenance item in order to get maximum service from the tyre. The pressure must be checked using an accurate tyre pressure gauge. The tyre is first inflated to a pressure sufficient to force the beads on to their seats. Many tyres have a line on the sidewall that should be concentric with the flange of the wheel if the bead is correctly seated. The pressure is now released from the tyre in order to allow the tube to take up a natural position, the valve core refitted and inflated to the proper working pressure. There may be a small loss in pressure some time after the tyre and tube have been installed, this being due to the escape of air that was trapped between the tube and the tyre.

Tyre pressure will vary with the load applied to the tyre and will increase by about 4% when the normal load is applied. Temperature change of 2°C is likely to cause a pressure change of about 1 p.s.i. (0.7 bar).

No mention has been made of tubeless tyres in the light aircraft situation. There was a relatively brief period when they were fitted as standard by most manufacturers of light aircraft but after several years a change was made back to the tube type of tyre. It is unlikely that any of these tyres are still in service. However there may be secondhand wheels around that were designed for use with tubeless tyres and are quite serviceable apart from the fact that it is not possible to obtain suitable tyres. Some wheel manufacturers did produce modification drawings that enabled these wheels to be adapted to use tube-type tyres. It is worth bearing in mind. However many of the heavier twins do fit tubeless tyres.

Following satisfactory assembly of the wheel and tyre, the bearings and dust seals may be fitted and the wheel refitted to the aircraft. The general procedure is to tighten the axle nut whilst slowly spinning the wheel. It should be tightened until a slight bearing drag is felt and then backed off to the nearest castellation that will align with the split pin hole in the axle. The legs of the split pin need to be bent over in such a way that they will not foul the bearing, the wheel hub or the dust cap. A number of aircraft manuals now specify a more refined method of tightening the axle nut. Typically the nut is tightened to a specified torque figure with the wheel rotating, in order to seat the bearing. The nut is then backed off and retightened to half the original torque value. Finally the nut is tightened further to permit alignment with the first cotter pin slot.

One aspect of wheel maintenance that has not been mentioned so far is wheel balancing. Whilst it is important that the balance marks on tyre and tube are lined up on assembly a small imbalance may still remain. On a low speed aircraft this may not notice but on those with higher take-off and landing speeds a vibration or nose wheel shimmy may be all too evident. Simple suspension balancers that will adapt to wheels used on typical aircraft under 5 700 kg maximum weight are available and are easy to use (fig.7.5). If the tyre and wheel assembly is in balance the bubble will be centred. If the bubble is not centred the wheel may be brought into balance with either rubber weights fitted inside the tyre or with lead weights on the wheel rim, depending on the type of wheel. A wheel may be equally well balanced on a set of knife edges with a suitable mandrel through the hub.

Tyres and tubes

There are many different styles, sizes and ratings of tyres and the choice appears bewildering. However one rule remains paramount - always fit the type, size and rating of tyre recommended by the aircraft manufacturer. There is no need to guess - it is in the Spare Parts Catalogue. Then again it is not wise to simply fit a tyre based on

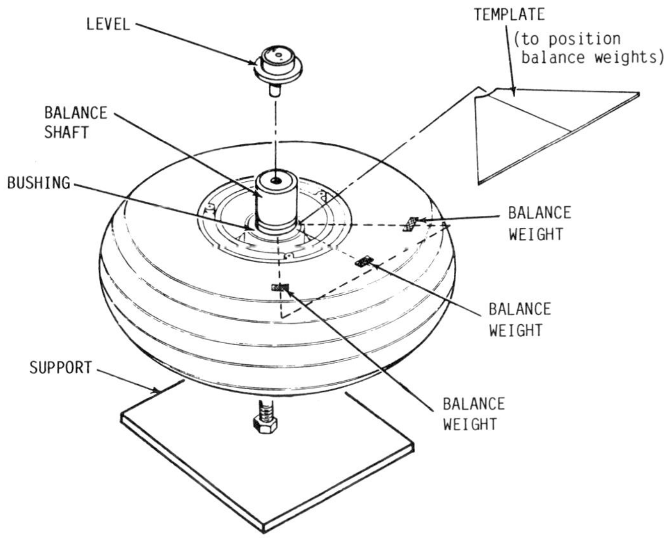

Fig. 7.5 Wheel balancer.

the one that was removed. The incorrect tyre may have been fitted on another occasion. A very popular light twin has experienced some growth over the years and has always retained the same tyre size, but to cater for the increased loading the ply rating was increased from 6 ply to 8 ply. Cases are on record of the incorrect rating tyres being fitted. It can be dangerous. So what do all the numbers mean?

Tyres are manufactured in different types for different applications but only three of these are commonly fitted on light aircraft today. The Type I – Smooth Contour (or Smooth Counter) is a tyre with the sidewall parallel to the outer surface of the aircraft wheel and is available in a variety of tread patterns. The most usual application is for helicopter handling wheels and for tail wheels. The Type I tyre is described by one dimension only, the outside diameter of the tyre. Thus a tyre described as 8.00 SC is 8 in. outside diameter.

Most main and nose wheel tyres are Type III tyres which are described as low pressure. These are measured by the width of the tyre and the inside diameter. Thus a 600-6 tyre will have a 6 in. width and a 6 in. inside diameter. An 800-6 tyre will be 8 in. wide but still has a 6 in. inside diameter.

The other type of tyre likely to be encountered is the Type VI, or low profile tyre. This type of tyre was developed to reduce the size of wheel wells and is currently fitted to several light aircraft with retractable landing gear. The outside diameter is added to the width and inside diameter when specifying this type of tyre. For example a 15–600–6 tyre is the same as a 600–6 tyre with the exception that the outside diameter is only 15 in. instead of the normal 16 in. It must be obvious that if the wrong tyre was fitted when a Type VI tyre was specified a total landing gear failure is likely to be the result.

Ply Rating is a figure that is based on the strength of the tyre carcass and is an indication of the maximum load that can be carried by the tyre. It must be remembered that when load is mentioned we are not only considering the static load imposed by an aircraft at rest but also the dynamic load imposed in landings that may be less than smooth. Typical ply ratings for light aircraft are 4, 6 and 8.

Tread designs vary with individual manufacturers and are usually designed for specific purposes. The most commonly used tread pattern is the ribbed design and this is suitable for most types of operation. With the different designs comes the question as to how far any particular design may be allowed to wear before it should be removed from service. Some tyre manufacturers provide recommendations in their service manuals, as may some aircraft manufacturers. However a general rule of thumb distilled from many sources would indicate that a tyre should be removed whilst there is still some evidence visible of the grooves nearest the centre of the tread. To run a tyre much beyond this will render it much more likely to failure through cuts and abrasions. Bear in mind that the cost of repairing the aircraft after an accident caused by a tyre failure is likely to far outweigh the amount saved by keeping the tyre in service as long as possible. In any case the legal minimum depth for a car tyre is 1 mm. and should we treat our aircraft to any less?

Tyre inspection

With the tyre mounted on the wheel there is relatively little inspection that can be carried out. Nevertheless it is important and should not be neglected. As has already been mentioned a check of the tyre pressure is the starting point for all tyre maintenance. When making visual inspections of the tyre remember the strength of the tyre is to be found in the carcass that is built up from the cords. The tread rubber provides a wearing surface and the sidewall rubber protects the carcass but neither provide the main strength of the tyre. For this reason the mass of small cracks often seen on the side wall of a tyre that has been in service for some time, and known as weather checking, is not detrimental to the strength of the tyre. A tyre with this type of defect may remain in service until such time as the cords are exposed.

Tread wear has been dealt with in general terms in the normal situation. Should more wear have taken place on one side of the tread than the other, a wheel alignment problem exists and the camber should be checked in the manner described in the previous chapter. In some aircraft the camber is not adjustable and the cause must be sought in some airframe or landing gear distortion, possibly caused by a heavy landing. Should the majority of the tread wear be at the edges of the tread pattern the usual cause is underinflation of the tyre. The tyre pressures quoted in the aircraft service manual may be for the average load and in any case should be regarded as a minimum pressure below which the aircraft should not be operated. Should the aircraft be operated close to its maximum weight most of the time and experience excessive tread shoulder wear the tyre pressure should be increased slightly. Take care not to exceed the maximum pressure for the tyre published by the tyre manufacturer. Since the early stages of abnormal tread wear are not easy to detect visually use an automotive tread depth gauge when making the inspections. If a tyre is starting to wear excessively on one side correct the cause and then fit the tyre the other way round on the wheel in order to equalise the wear. Excessive tread wear in the centre of the pattern is caused by overinflation and tyre pressures should be adjusted to take account of the normal operating weight of the aircraft.

Cuts in the tread of the tyre may not be detrimental to the strength of the tyre and are usually caused by foreign objects on runways and taxying areas. Small chevron-shaped cuts across the width of the tread are caused by running and braking on cross-grooved runways. If a cut extends more than 50% across any rib, or if the cords in the carcass are exposed, the tyre should be scrapped. Turning on the spot using the brakes is likely to put a flat spot on the tyre which, even if it does not scrap the tyre by removing the rubber, will put the tyre out of balance. An out of balance tyre will tend to develop undulations in the tread pattern round the circumference, the severity depending on the severity of the imbalance.

External inspection can only show the faults associated with the wearing surfaces of the tyre and tells relatively little about the condition of the carcass underneath. In the event, for example, of a heavy landing the tyres should always be removed for internal inspection since there can be complete failure of some of the cords which will not be revealed by an external examination. Cord failure may sometimes cause a bulge on the tyre and such a symptom is cause for tyre removal and usually scrapping. When a tyre is to be inspected internally it should first be examined for any signs of deteriorated rubber, particularly around the bead area, which would indicate excessive heat from brake application. To properly inspect the inside of the tread and wall areas the tyre should be rolled slowly along the bench or the floor whilst exerting a downward pressure. This will cause the tread area to flatten out and show up any cord damage that may have occurred. A powerful torch, or other light, will be necessary when making this inspection. Typically a concussion failure, that is a failure through impact with a sharp object, will show up as a cross-shaped fracture in the inner casing. Any of these conditions is a reason for scrapping the tyre.

Generally speaking the tube should be replaced at any time that the tyre is replaced, although this requirement may be waived if the time in service is relatively short and the tyre has not run flat. If puncture repairs are made they should only be made with a vulcanising type of patch and limited to one per tube. To check for leaks the tube may be inflated and immersed in water and at the same time a visual examination made for any surface cracks that may be revealed as the rubber stretches under the pressure. With the wheel on the aircraft valve core leakage may be detected by turning the wheel until the valve is at the highest point and placing a small container of water over the stem. A valve core is cheap enough and should be replaced using standard tools any time one is suspect.

Brake units

Light aircraft brake units may be readily divided into drum and disc types. The disc brakes may further be subdivided into fixed and floating single disc types and the multi-disc types. The multi-disc brake is not usually found on single-engined aircraft and tends to be confined to the larger twins. The drum brake was extensively used at one time either in the form of an automotive type shoe brake or as an expander tube type of brake. In the latter the hydraulic fluid under pressure is fed into a flat tube which expands and forces the brake blocks out radially into contact with the drum. However current production aircraft use the disc brake in one form or another. The disc brake is always hydraulically operated but there have been mechanically operated drum brakes.

Brake inspection and repair

The order of dismantling will vary with the design of the brake; for example, with a floating disc brake (figs.7.6 and 7.7) it will be necessary to remove the wheel before the brake unit but in a fixed disc brake (figs.7.1 and 7.2) the brake unit can be removed independently of the wheel. It will also be necessary to disconnect the

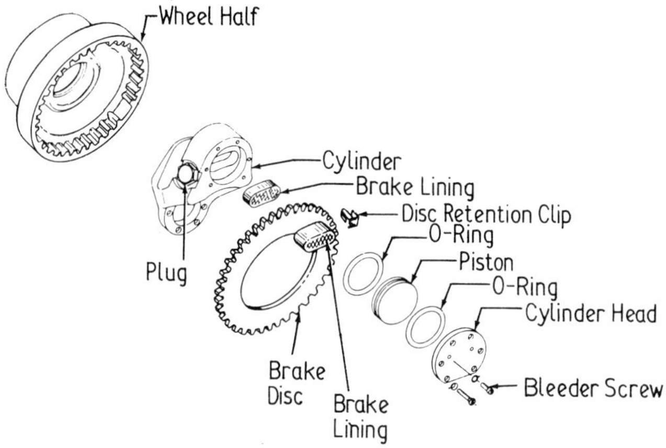

Fig. 7.6 Floating disc brake.

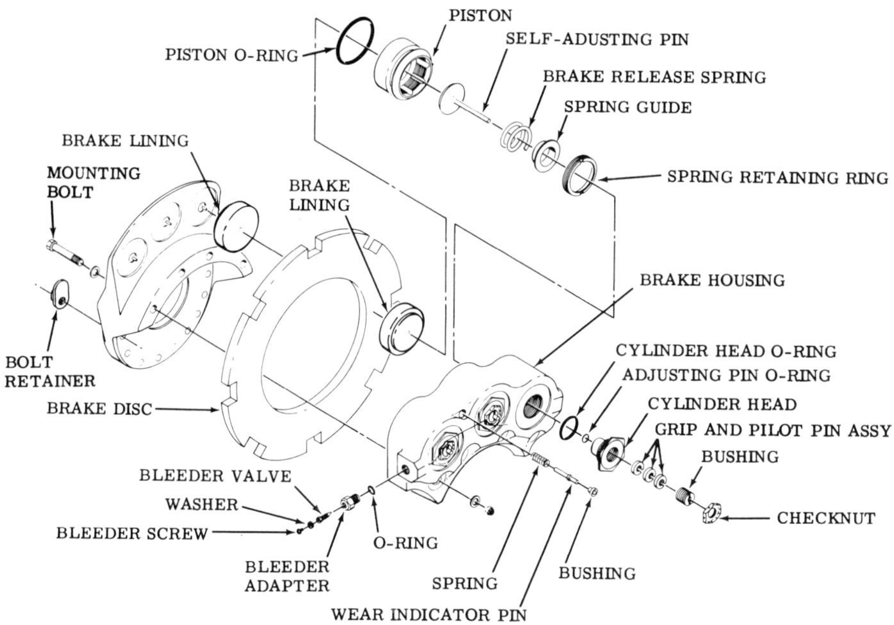

Fig. 7.7 Multi-piston brake.

hydraulic line from the brake unit and to catch the fluid spillage in a small can. The open end of the brake line should be blanked as soon as possible after removal in order to prevent the ingress of any dirt or other contamination. Masking tape is not a suitable means of blanking the line since the hydraulic oil will soften the adhesive and contaminate the system and only proper metal or plastic blanks should be used. Cleanliness in a hydraulic system is next to serviceability. Before stripping the removed brake unit, thoroughly clean the outside of the parts with a suitable solvent.

The pistons may be removed from the hydraulic unit by placing a shop air line against the inlet fitting and giving a quick blast of air. It will be necessary to ensure that the pistons are restrained so that they do not fly across the workshop and cause damage. The O-rings may be removed from the pistons using any suitable non-metallic tool, such as a plastic rod filed to a flat end rather like a small screwdriver. A metal tool would damage the piston. All the parts removed should be cleaned again in solvent and thoroughly dried and spread out on a clean metal-topped bench. If the job is left at any time the clean parts should be covered with a piece of polythene. The bore of the cylinder and the circumference of the piston should be inspected for scores and other damage.

Fine scoring should be polished out using 600 wet and dry paper, preferably with a little kerosene as a lubricant. More severe scoring is cause for rejection of the affected parts. The bleeder nipple should be thoroughly cleaned and the passages flushed through with solvent. Before reassembly all parts should be given a final rinse in clean solvent and dried. The piston and O-ring should be lubricated in the appropriate hydraulic fluid and the ring gently stretched and fitted into its groove. The cylinder bore should be lubricated in the same way and the piston carefully offered up, taking especial care that it does not tip to one side or the other and jam. It should be pushed fully into the cylinder. The cylinder may now be refitted to the aircraft and the brake hydraulic system bled of air.

The brake friction pads may be in the form of a kidney-shaped block of material and replacement is straightforward and obvious. In other designs of brake the friction pads are in the form of flat linings made from a brass-reinforced fibre material. An improvement of the design used on heavier aircraft uses linings that are made of sintered metal. The maximum permitted wear on flat linings is usually specified in the form of a minimum thickness for the lining and this is easily measured in service by using a drill shank of the appropriate size as a comparison gauge. In the first type of brake assembly, using kidney-shaped pads, the wear allowance is usually specified in the form of a maximum (or minimum) clearance between the floating disc and the part of the brake housing containing the stationary lining. A drill shank may be used in a similar manner as a feeler gauge.

Flat type fibre linings are retained in place on the pressure and back plates with semi-pierced brass rivets which may be removed by drilling with a slightly undersize drill until the head of the rivet breaks free. The shank of the rivet may be driven out using a parallel punch whilst the pressure plate is supported on a hollow mandrel. The replacement rivets in the new linings are spread using a specially shaped punch designed for the job. The sintered metal linings are simply prised away from the pressure plate using a thin screwdriver or a knife. The replacement linings snap into place although it may be necessary to gently squeeze the assembly in a vice with soft jaws to make sure that they are properly seated.

Once new linings have been fitted to the brake it is important that they are properly conditioned before the aircraft is released for service. Failure to do so will result in reduced braking efficiency and shortened lining life. In the case of organic (fibre) linings the conditioning process cures the resins in the material. If excessive heat is applied before curing, for example by high energy braking, the lining material will be carburised, thus preventing proper braking efficiency. The correct conditioning procedure is to carry out a minimum of 6 light brake applications from about 25 to 40 m.p.h. Allow the brake discs to partially cool between stops. Metallic linings require conditioning in order to glaze the surface. An unglazed lining wears quickly and can severely score the discs. The linings are conditioned by performing three consecutive hard braking applications from 45 to 50 m.p.h. Do not allow the brake discs to cool substantially between stops. Once the linings are conditioned the braking system will provide many hours of maintenance-free service.

Brake discs and drums wear and should be inspected and repaired at the same time as the brake unit. Brake discs can stay in service until the minimum thickness specified by the manufacturer is reached. There may be a repair scheme that will allow discs to be skimmed in a lathe but this should never be done unless proper data are available. But thickness is not the only criterion when inspecting brake discs. Any disc is liable to wear unevenly and score as it collects dirt and grit from runways. Mild scoring is not cause to reject a disc but severe scoring means automatic scrap. Likewise a disc should be checked to see that it is running true and is not warped. This may be readily achieved by placing the disc flat on a surface plate and checking round the inner and outer circumferences with a feeler gauge. The key slots in a floating disc should be checked against the manufacturer's tolerances. The welds in a fixed disc should be checked for cracks. A brake drum should be inspected in a similar way to a disc but the critical dimension is that of inside diameter.

The multi-disc brake that is being increasingly used on the heavier twin-engined aircraft employs two rotating discs that are keyed into and rotate with the wheel

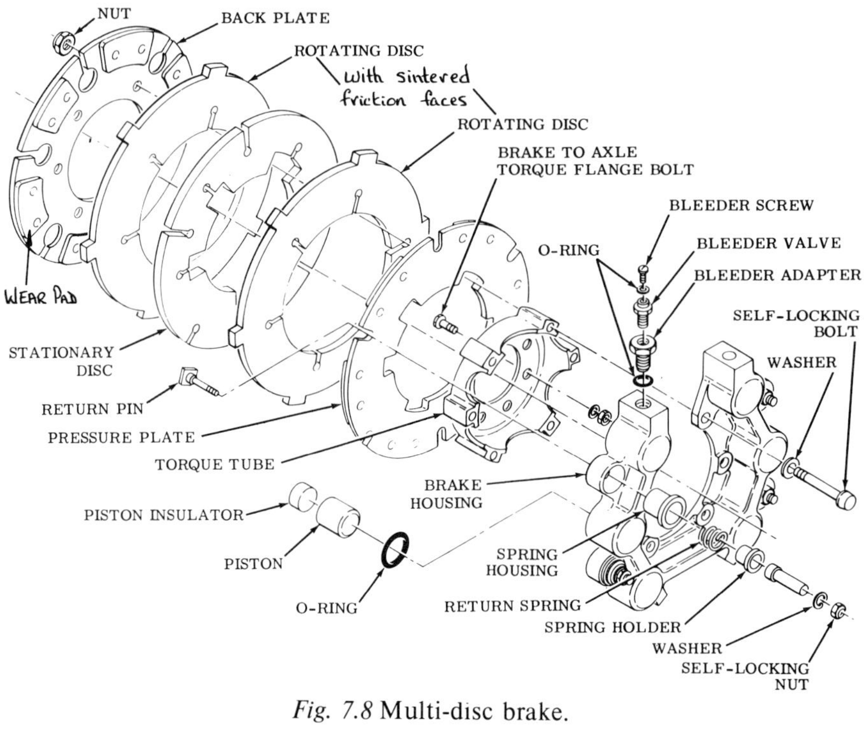

Fig. 7.8 Multi-disc brake.

(fig.7.8). The rotating discs are sandwiched between three stationary discs and pressure is applied by five or six hydraulically operated pistons. The stationary discs are either plain or fitted with wear pads and the rotating discs are coated with a sintered metal to provide the friction material. Brake wear on this type of brake is measured by applying the brake and measuring the clearance between the brake housing and the pressure plate. As brake lining wear takes place so this distance will increase until the maximum allowable is reached. Inspection of the individual parts is similar to any other disc and lining, with minimum thicknesses and key dimensions being quoted in the service manual. The sintered linings cannot be replaced and the entire disc must be replaced if it is worn below the minimum thickness. The pressure and back plates have renewable steel wear pads which are riveted to the discs and may be replaced. After they have been renewed it is necessary to surface grind the pads to the required thickness and therefore the smaller workshop would not be able to handle this type of repair.

Hydraulic brake systems

Irrespective of the style of hardware installed at the wheel a system capable of applying sufficient pressure to the pistons and capable of being operated by the pilot is required. Whilst there are one or two exceptions to the rule most aircraft use hydraulic master cylinders with a separate system to each wheel brake. One manufacturer of heavier twins uses fully power-operated brakes with the main hydraulic system of the aircraft providing the pressure source. Among aircraft using independent master cylinders there are two basic variations, those with independent reservoirs for each wheel system and those using a common reservoir. This tends to be a matter of choice among manufacturers, there being no real technical advantage for either system. It is easier to change the seals in the former and the latter is easier to check and top up the fluid. The description that follows is applicable to any basic type of master cylinder although there may be minor variations in construction.

Master cylinders

In use the master cylinder piston rod is forced downwards by foot pressure on the upper part of the rudder pedal (fig.7.9). As this happens the Lock-O-Seal contacts the upper face of the piston sealing off the compensating orifice and any further movement moves the piston down compressing the hydraulic fluid in the system below it. On the return stroke the piston and piston rod rise together until the piston contacts the compensating sleeve. Further movement of the rod separates the piston from the Lock-O-Seal and hydraulic fluid is able to flow through the ports in the compensating sleeve and through the centre of the piston to make good any losses in the wheel brake system. The correct operation of this type of master cylinder depends on there being a specified clearance between the Lock-O-Seal and the piston, with the piston in the released position. Typically this clearance is of the order of 0.040 in. and is not easy to set since, with a large clearance hole in the centre, the piston tends to tilt if feeler gauges are used. The job is made easier if a special gauge is made from a twist of standard American 0.040 in. safety wire as shown in fig.7.10. The two legs of the gauge should be far enough apart to sit on the washer surrounding the seal. It should be noted that the filler plug is drilled with a vent hole and blockage of this hole will cause erratic operation of the brakes. Should there be little resistance to the operation of the brake pedal the most common cause is failure of the Lock-O-

Seal. After replacing the seal the top of the piston may be smoothed on a fine oil stone using kerosene as a lubricant and taking care to remove all burrs and sharp edges, especially round the centre hole. One advantage of this type of cylinder is that all such work may be carried out without draining the brake system.

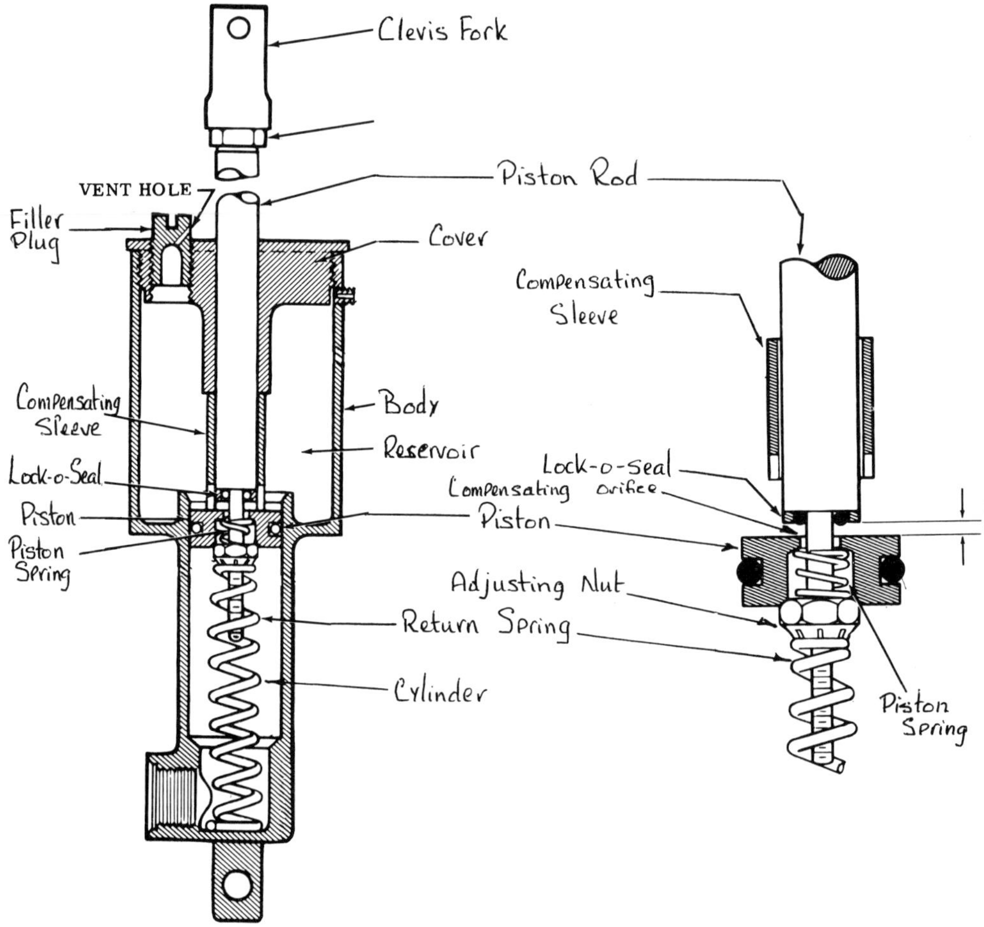

Fig. 7.9 Hydraulic master cylinder.

Fig. 7.10 Clearance gauge.

Bleeding the brake system

Whilst it is possible to bleed the air from the brake system by filling the reservoir with hydraulic fluid and pumping away until it all works its way through to the brake unit, it is not a very efficient way to do the job. Better by far is the pressure method which fills and bleeds the system in one operation from the bottom. Assuming that the system is empty a pressure can of hydraulic fluid is connected to the already slackened bleeder screw on the wheel brake unit. A normal pump action oil can kept specially for hydraulic fluid will be quite satisfactory. The hydraulic fluid is pumped into the system until the reservoir is full, having expelled the air in the process, and the bleeder nipple is tightened before removing the pressure source. If this method is being used with a system having a common reservoir then the first system should be filled sufficiently to bring the level to just cover the bottom of the reservoir. This will allow sufficient room to bleed the second system properly without the danger of any air being trapped in it.

Hydraulic fluids

A word about the different types of hydraulic fluids is in order at this point. Whilst most light aircraft use aircraft hydraulic fluid in their brake systems never take it for granted. There are one or two aircraft that actually use automotive brake fluid. The significance of this fact is that it is necessary to use seals of a different material and the use of the wrong fluid will destroy the seals. This may not be evident at first but could easily lead to an entry in the accident statistics at a later date. So check the service manual and make sure which fluid is required. To confuse the issue a little further there are a couple of aircraft that use automotive brake master cylinders but not with the original seals! Do check the parts catalogue for the correct part before fitting it.

Given a little care and attention the modern hydraulic brake system does a remarkably good job with little fuss. Bear in mind, however, that the amount of maintenance required will depend upon how the aircraft is handled and the type of surfaces from which it operates.

CHAPTER 8
Flying Control Systems

Since the early days of flying a vast amount of work has been put into the design of flying control systems in order to provide positive and responsive control of the aircraft throughout the flight regime. Control at low speeds makes different demands to those made by control at high speeds and the final result is often one of compromise. The control systems of light aircraft can be divided into three groups:

1 Primary flying controls, i.e. ailerons, elevator, rudder, stabilator.
2 Auxiliary flying controls, i.e. trimmers and trim tabs.
3 High lift devices, i.e. flaps, slats.

This chapter will be confined to manually operated controls with the exception of flap systems. The actual hardware used in control systems is common to fixed-wing aircraft and rotorcraft although the latter systems also contain some items not found in other aircraft. Rotorcraft control systems will be discussed specifically at the end of this chapter but the majority of rigging procedures are valid for all types of aircraft.

Basic layout - primary flying controls

The primary function of flying controls is to translate pilot demand into changes of aircraft attitude and hence flight path. Although the primary flying controls are fairly standard among light aircraft the layout of each system varies according to the design of the aircraft and the manufacturer's own preferences. The typical systems shown inevitably reflect only a small percentage of the possible variations.

In the aileron system shown (figs.8.1 and 8.2) the control wheels are linked together by a chain and sprocket system. Turnbuckles provide a means of achieving the correct chain tension and alignment of the control wheels. The chain system rotates a bellcrank which, in turn, operates the flexible cables passing to the aileron bellcranks in the wings. These bellcranks are linked together by a balance cable in order to maintain the tension in the cable system which is adjusted by turnbuckles. The movement is finally transmitted to the ailerons via push-pull rods. The stops limiting the range of movement are to be found in the form of bushings within kidney-shaped slots in the bellcranks. Other types of aircraft utilise adjustable stop bolts in the wing structure that contact the bellcrank at the extremes of movement.

The cable-operated elevator system (fig.8.3) is even simpler than the aileron system in that two bellcranks at each end of the system are joined by cables. The input to the forward bellcrank and the output from the aft bellcrank are by push-pull rods. Stops are provided at the control surface or the bellcrank to limit the range of movement.

The rod-operated elevator system substitutes push-pull rods for the cable system. Since a push-pull rod decreases in rigidity with increase in length it is necessary to limit the length of the individual rods in the system and this leads to the use of idler bellcranks and support rollers.

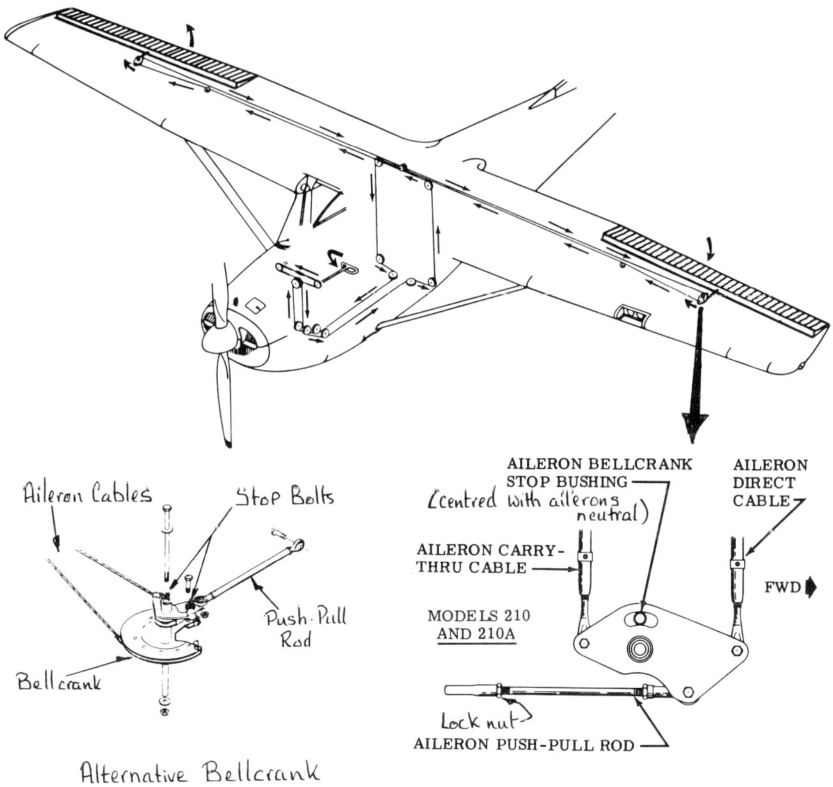

Fig. 8.1 Aileron control system.

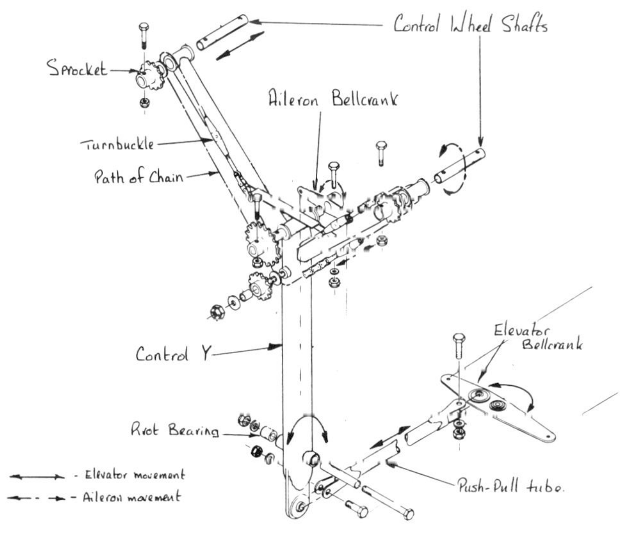

Fig. 8.2 Control 'Y'.

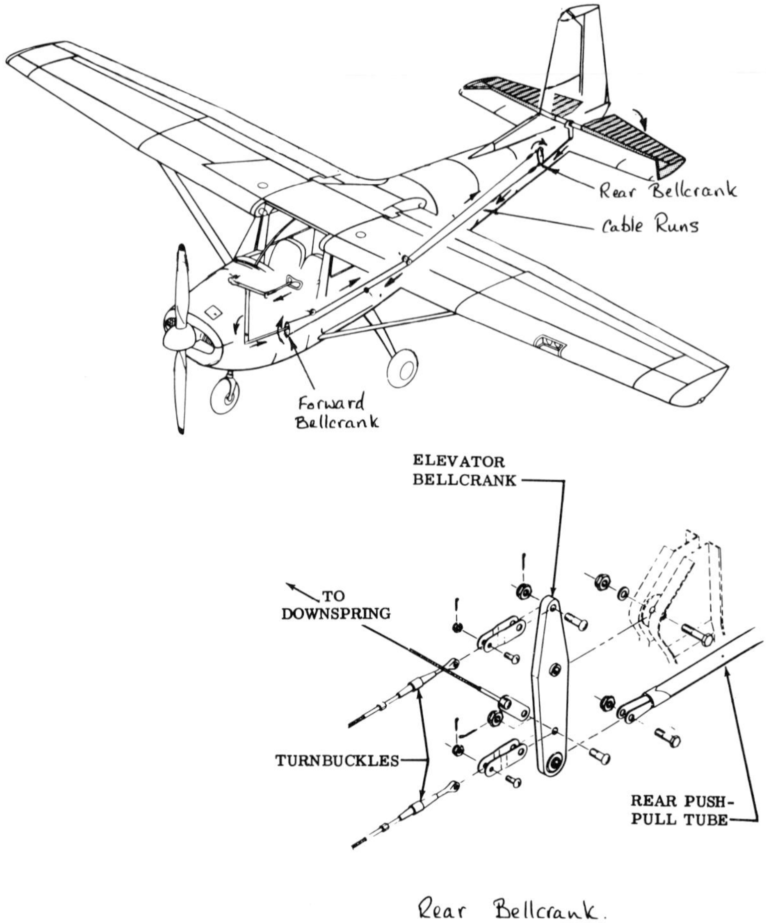

Fig. 8.3 Elevator system.

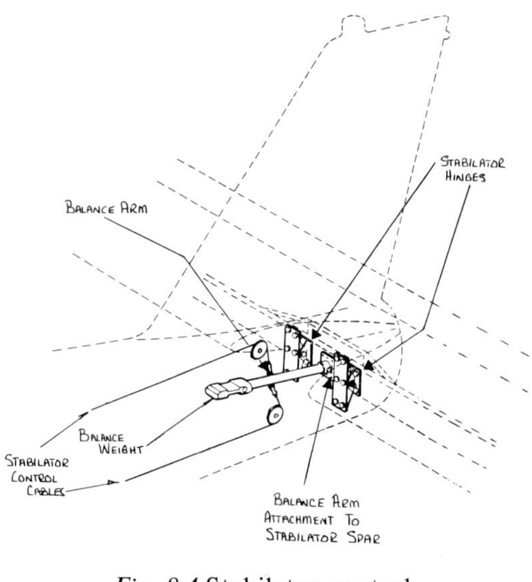

Fig. 8.4 Stabilator control.

The stabilator, or 'all flying tail', operates in a similar manner to the elevator system for which it is an alternative (fig.8.4). The primary difference lies in the connection of the control at the aft end. The stabilator system uses a long arm inside the fuselage to provide the mass balance and the control cables are connected directly to this arm.

Rudder systems are intrinsically simple and follow the basic layout for any cable-operated system. The rudder pedals may vary greatly in design although with three main variations. In some aircraft parallel torque tubes across the floor join left and

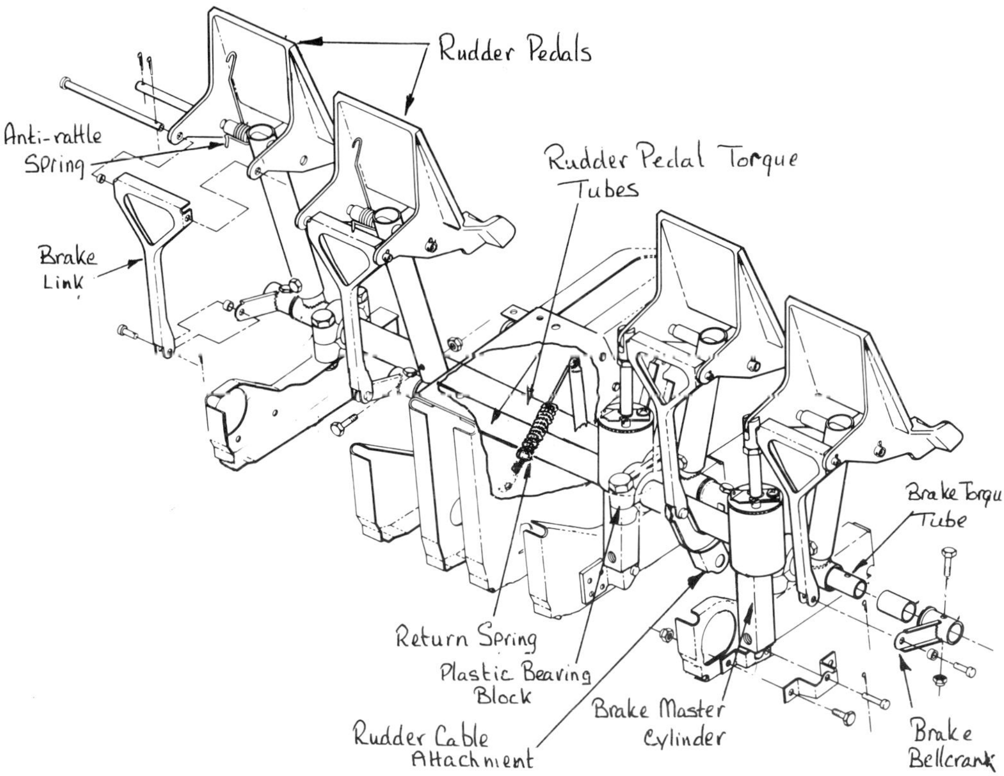

Fig. 8.5 Rudder pedals.

right-hand sets of pedals (fig.8.5) whilst in other types the pedals are suspended from the torque tubes. The movements of the pedals to operate the brakes are mechanically interconnected in either case using concentric torque tubes and levers. Alternatively pedals may be mounted on a rudder bar in which case interconnection of the brakes is often accomplished hydraulically (fig.8.6).

Basic layout - auxiliary flying controls

All types of trim tabs operate in a similar manner so only the elevator tab has been illustrated. The intermediate part of the system is typical of any cable-operated system but the basic differences between the two systems illustrated lie in the methods of driving the cable and in the type of stops. In fig.8.7 the cable is joined to a short length of chain at each end which provides a positive connection to both

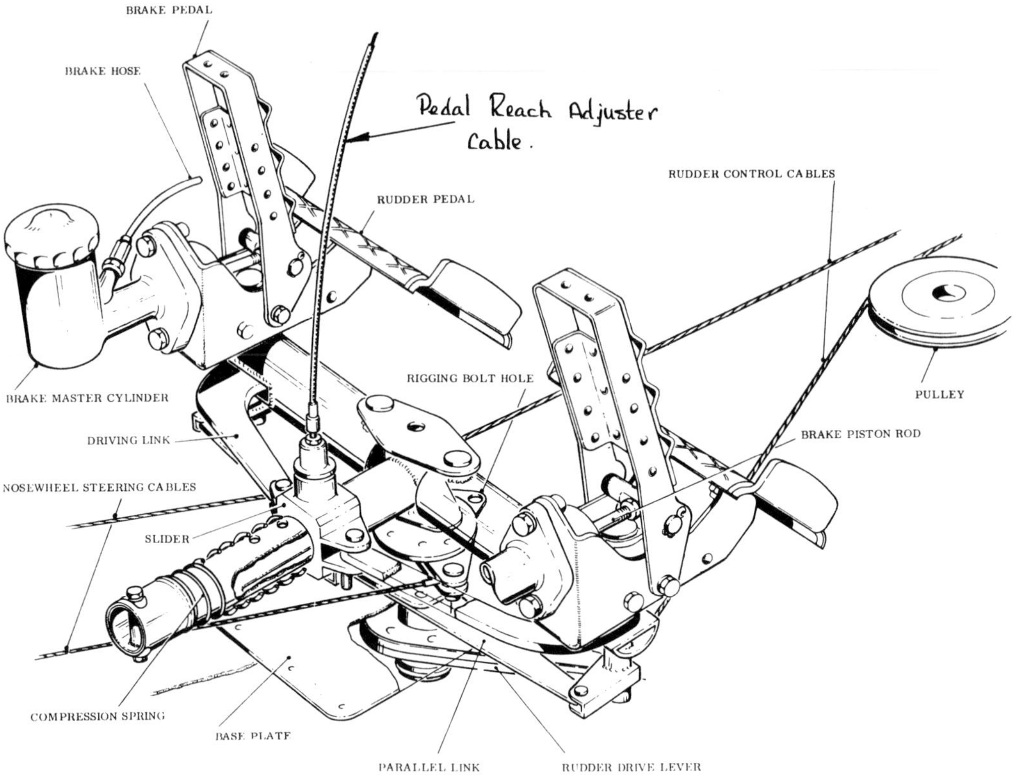

Fig. 8.6 Rudder bar assembly.

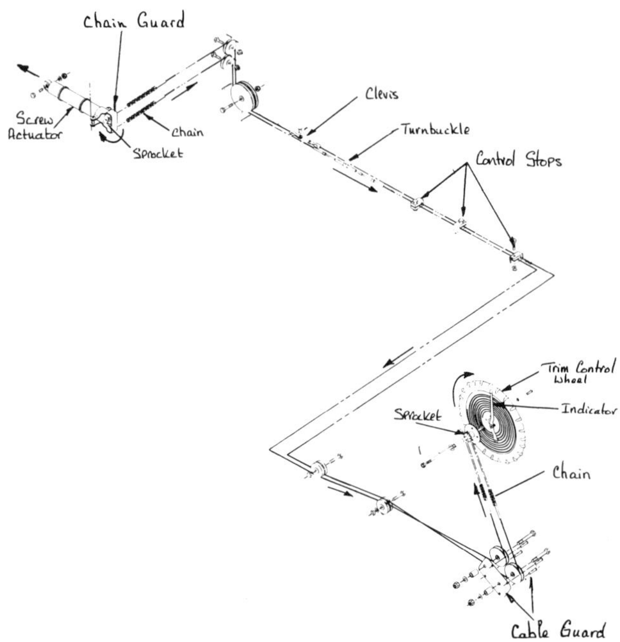

Fig. 8.7 Elevator trim system.

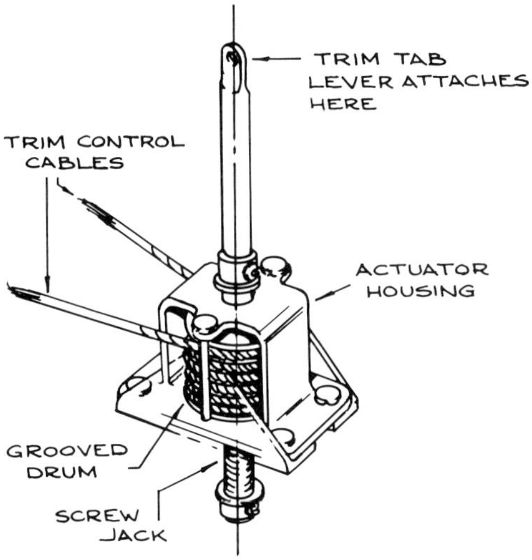

Fig. 8.8 Stabilator trim jack.

handwheel and operating jack. In fig.8.8 the same ideal is achieved by swaging a ball on to the cable at predetermined locations at each end. This ball fits into a slot in the cable drum to provide the positive connection. In the former system the travel is limited by stops clamped on to the trim cables whilst in the latter the stops form part of the screw jack. A further difference lies in the indicator systems used. One uses a wire pointer operated by a spiral groove in the hand wheel whilst the other transmits tab position by means of a thin piano wire to a spring-loaded pointer in the cabin.

There are some aircraft that do not actually use a trim tab to relieve control loads but operation of the trim control applies spring pressure in the desired direction. This type of trim system is often known as a spring bias. However a heavy spring also forms part of the elevator system of some modern aircraft in order to increase control loads artificially as the pilot moves the control towards nose up. This is employed as a safety device in an aircraft where there is insufficient natural aerodynamic warning of this movement.

Control system sense is best visualised by imagining the cockpit control as the old style stick when the control surface moves to meet the stick. Move the stick to the right and the right aileron comes up to meet it; move it aft and the elevator comes up to meet it. Trim tab movement will always be in the opposite direction to the main surface to achieve the same change in the flight path of the aeroplane. In the case of the rudder, left pedal forward will move the rudder to the left.

Control system hardware

Whatever the variations in layout of the control systems there is a commonality in the hardware used. This section cannot be a directory of the different types available nor of the differences between various national standards but it will attempt to describe and discuss the items in general terms. With regard to inspection of cables and other components much has been written in the companion volume to this book, 'Light Aircraft Inspection', to which the reader is referred.

The basic element in the majority of control systems is the flexible cable. The flexible cable consists of a number of steel wires twisted, or preformed, into strands;

the strands are similarly twisted, or preformed, into a cable. The term preformed is applied to a cable where the strands are laid in such a way that they are, in effect, shaped to their position in the cable. The strands of a preformed cable will stay in place if the cable is cut whereas those of a simple cable will immediately unravel. The wires may be of carbon steel or of stainless steel, the latter being particularly favoured for marine aircraft. During the manufacture of the cable a lubricant is applied that stays in the cable to reduce internal wear and to improve the fatigue life, and for this reason cables should never be cleaned with solvents. A wipe with a rag should be sufficient to remove accumulated dirt. Non-stainless cables are often protected with a lanolin-based compound to exclude moisture and reduce the risk of corrosion, such compound being identifiable by the fact that it is coloured. Cables are sized by diameter in the U.S.A. and care must be taken to rotate the cable when measuring to ensure that the maximum diameter is being measured. In the U.K. cables are sized by breaking load in hundredweights (1 cwt = 112 lb).

Inspection may require partial removal of the cable from the aircraft in order to gain access to portions that would be otherwise hidden. This is facilitated by tying a piece of string to the end of the cable before withdrawing it. It will now be a simple matter to replace it in its correct position in the aircraft structure. Broken strands may be readily found by drawing a rag along the length of the cable. At locations where a cable flexes in operation, for example round a pulley, wear will take place internally and may lead to failure of internal wires. To find such failures the cable should be bent in a curve as it is passed from one hand to the other, whereupon broken internal wires will emerge at the surface of the cable. A visual inspection of the cable should be made for corrosion or flattening of the strands caused by external wear at fairleads and the like. If the cable has been subjected to a sudden tensile load that was insufficient to break the cable but was near its upper limit, the preformed strands may straighten out at the point of maximum stress. When the tension is removed from the cable these strands will no longer conform to the shape of the cable, giving rise to the effect known as 'bird-caging'. Cables may also kink but both defects may be found during the course of the normal visual inspection.

The ends of the cable require to be terminated in a manner that will permit them to be connected into the system (fig.8.9). The simplest form is a swaged sleeve used in conjunction with a cable thimble. The accuracy of the swaging is checked by measuring the major dimension of the swaged portion with a go-no go type of gauge. In use the thimble is attached to the bellcrank or lever with a small shackle or directly with a bolt. Alternatively threaded cable terminals may be swaged on to the cable and these mate directly with other fittings such as turnbuckles. Cable inspection should always include the swaging for cracks and slippage. Cracks are, perhaps, more likely at the time of the swaging process although slippage could take place at any time. Slippage would be evidenced by a shiny portion of the cable adjacent to the fitting or by breakage of the paint seal applied at the ends of the swaging. In the case of a swaged ball terminal the cable must protrude through the fitting for it to be satisfactory.

Turnbuckles are used in cable systems to make minor alterations to the length of a cable and to provide a means of adjusting the cable tension when installed. The central barrel is threaded at both ends, one being right hand and the other left hand, thus rotation of the barrel will lengthen or shorten the assembly as required. American turnbuckles are usually marked with a circumferential groove to indicate the left-hand thread. To ensure that sufficient thread remains engaged in the barrel

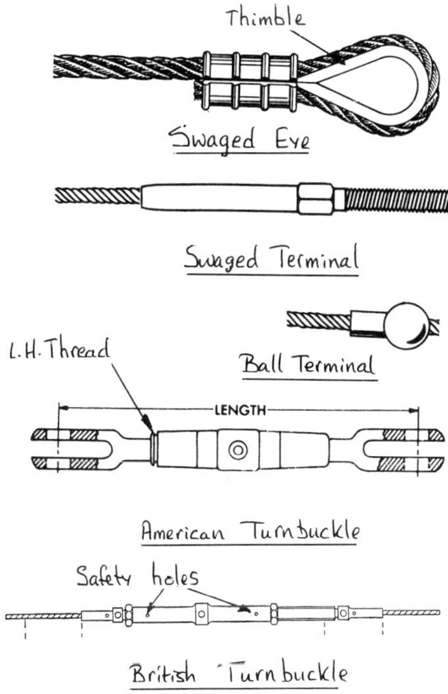

Fig. 8.9 Cable hardware.

various safety devices are used. With a U.S. turnbuckle there must be no more than three threads showing at each end of the assembly. In a U.K. turnbuckle small holes are drilled in the barrel and it must not be possible to push a piece of steel wire through the holes. Steel locking wire is often used for this purpose but it is rather flexible. A better tool is made by unfolding a wire-type paper clip. If no safety hole is provided on a U.K. turnbuckle no threads may be showing at the ends for a safe condition. Sometimes turnbuckles will be found that utilise a rod as the central element instead of a barrel and in this case the safety holes will be found in the rod ends. This type is locked against unscrewing by means of lock-nuts. The barrel type turnbuckle is locked with steel lock wire in the manner shown in fig.8.10. Although a single locking is permissible the double, or figure-of-eight, locking is preferred. An alternative to wire locking is found in the clip-locking turnbuckle. After adjustment the end fittings are rotated so that machined grooves align with matching grooves in the thread of the barrel. A spring steel clip is slid into position in the groove and its free end is pushed into the central hole in the barrel. Locking wire and locking clips must not be reused. If locking clips are not available the turnbuckle may be locked with wire in the normal manner.

Cable direction is changed by means of pulleys which are commonly made from one of the resin/fabric materials or may be moulded from a thermo-setting plastic. The pulley may rotate on a plain bearing but more often on a ball bearing. A pulley with a plain bearing usually requires a thin washer on each side and the retention bolt must be tightened enough to remove excessive side play and yet leave the pulley free to rotate. A self-locking nut is not acceptable in this type of installation. The bearing bush should be lubricated prior to assembly. Where a pulley rotates on a ball bearing it is lubricated for life and the mounting bolt should be tight in order to nip the inner

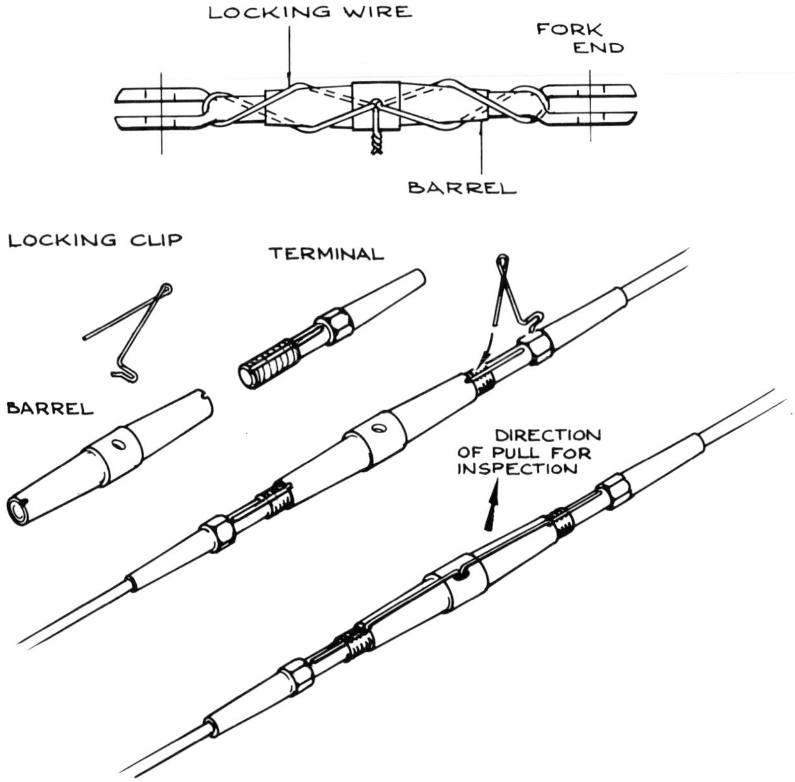

Fig. 8.10 Turnbuckle locking.

race of the bearing securely. Failure to do this will result in worn holes in the mounting structure. Periodic inspection of pulleys should verify that the groove is

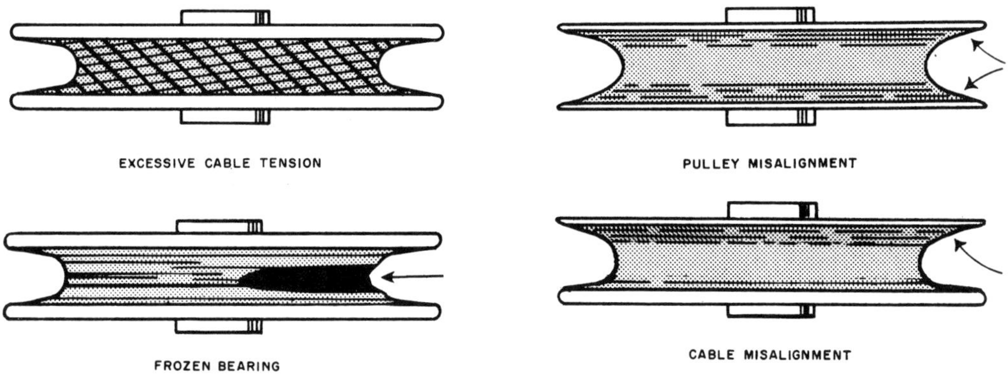

Fig. 8.11 Pulley wear.

not excessively worn and that the flanges are not damaged (fig.8.11). Light marking of the bottom of the groove is acceptable but the pulley should be rejected when this becomes heavy. Misalignment of the cable will cause one or both flanges to wear thin. The pulley should be free to rotate. Most pulleys using ball bearings use sealed bearings and these cannot be relubricated. To prevent a cable jumping out of the pulley groove a guard is usually fitted and may be in the form of a piece of sheet metal secured by the same bolt as the pulley or may be a split pin through the mounting structure. Make sure that they are in place, cannot rotate and do not foul the cable.

Where cables pass through, or close to, aircraft structure fairleads made of nylon or fibre are used. These may be tubular or flat as is most suited to the location. A fair amount of wear may be allowed as long as the fairlead is likely to continue its correct function until the next time it is inspected.

Cables are often attached to quadrants or bellcranks so that their linear motion may be converted to rotary motion or in order to change the ratio or direction of motion. Cables may be attached in a number of ways and must always be fitted in the way shown in the manufacturer's manuals. On a quadrant, where the cable runs in a circumferential groove, a ball end swaged on the cable may be located in a slot and retained by a clevis pin safetied with a split pin. In other cases the cable terminal is held in a fork end with a bolt and nut or attached with a shackle. All these parts should be periodically inspected for wear at the points where the cable tension is constantly felt. Check cables and terminals for free movement in relation to the bellcranks. Loss of this free movement will cause the fittings or cable to flex in a way for which they were not designed and will result in a fatigue failure. The bellcranks themselves pivot on bushes or bearings and these require periodic checking for wear and lubrication. If fork ends on bellcranks and fittings, especially at aluminium alloy bellcranks, are overtightened they are likely to crack at the change of cross-section at the base of the fork. Where excessive clearance exists on assembly it should be taken up with standard washers prior to tightening the bolt. Such shimming should be disposed equally on both sides of the fitting and a check should be made for full and free movement in all positions.

Movement of the bellcrank is often transmitted to the control surface via a push-pull rod which may be fixed or adjustable. In applications with low loads and a short distance the assembly often consists of a solid rod, threaded at the ends, and two standard rod ends with self-aligning spherical bearings. These are locked in position with lock nuts. Where rods are longer and loads are higher a tube is usually used because of the greater rigidity in compression for a given weight and strength. Fittings riveted into the ends allow attachment of rod ends in the same manner as the smaller rod. When a push-pull rod is being adjusted for length when rigging a control system it is important that the bearing is held with a spanner whilst the lock nut is tightened (fig.8.12). If this is not carried out correctly the bearing race will be damaged causing excessive friction and shortening its life. All rods should be inspected for chafing on the aircraft structure and for cracks and corrosion. Cracks are most likely at a change of cross-sectional area, for example, at the last thread on a solid rod. Corrosion is most likely in the inside of hollow rods, especially if they are vertical, and suitable internal inspection must be carried out periodically. Since water can creep along the threads a silicone-type RTV sealing compound should be applied over the end of the tube and the lock-nut. All surfaces should be thoroughly cleaned before applying the compound. Such a compound is easily removed with a sharp knife when inspection or adjustment is necessary. Rod ends should be checked for wear, bearing in mind that backlash in a control system can cause control surface flutter and all kinds of rotor induced vibrations. Lubrication may or may not be required for the spherical bearings and is probably best omitted under dry and dusty conditions in order to avoid dirt collecting in the bearing. Teflon-lined bearings must not be lubricated under any circumstances, but they should be inspected regularly

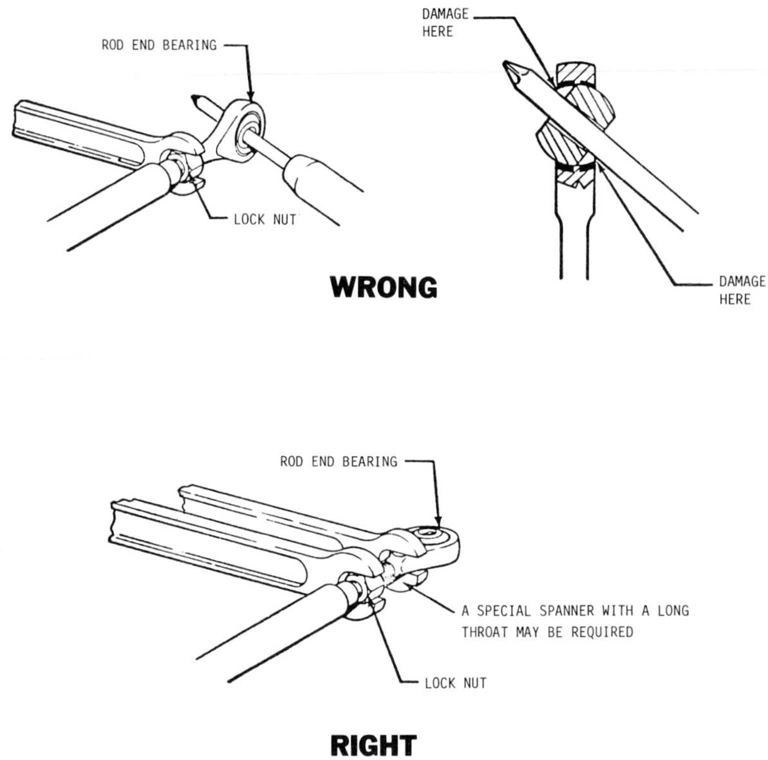

Fig. 8.12 Rod end bearing adjustment.

since once a small amount of wear has taken place the clearance increases rapidly to the point where the teflon breaks up.

Small changes of direction and alignment in control shafts and torque tubes are often accomplished by using a universal joint. The majority of these are simple and little is required other than routine inspection. Free play should be assessed and compared with a new item. Some joints in use have a considerable amount of backlash even when new and this should be taken into consideration when making the inspection.

Frequently rotary motion is transferred to and from cable systems by the use of chains connected to the ends of the cables. This is particularly true in trim systems. The chains are simple roller chains used in conjunction with sprockets on the drive shafts. Chain assemblies are attached to the cable fittings with bolts or clevis pins and the use of spring clip connection links is not allowed. In some instances dual-control wheels are linked and synchronised using chains, and turnbuckles are provided to permit adjustment of the alignment of the wheels. Where a continuous chain system is used an adjustable idler pulley provides the means of adjusting chain tension. Guards are provided at sprockets for the same purpose, and in a similar manner, to those provided for pulleys. Inspection of chain assemblies should start with a check for smooth movement of the chain over its full range of movement. The chain itself should be inspected for cleanliness, adequate lubrication, excessive wear and corrosion. Excessive wear should be checked by degreasing the chain, laying it on a flat surface and extending it under a small tensile load. A typical load would be 16 lb for a chain with a pitch of 0.375 in. The length of the chain under this tension should not exceed 2% more than its original nominal length. The original nominal length is calculated by measuring the pitch and multiplying it by the number of pitches. It is important to relubricate the chain immediately the check is finished if it is to be reused in order to avoid corrosion starting. The shape of the teeth on a sprocket should be checked against a new one, the wear usually taking place at the base of the teeth, thus making them hook-shaped.

Screw jacks are frequently used in trim and flap operating systems where large, rapid control movements are not required (fig.8.13). Typically a trim tab screw

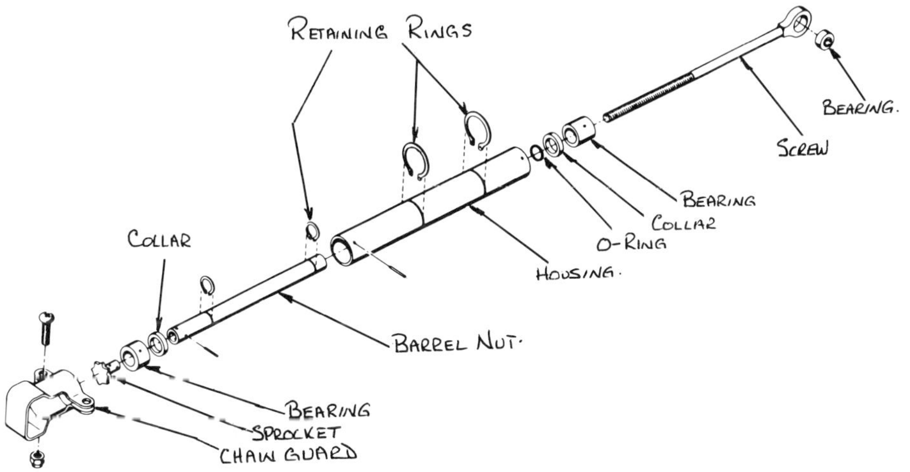

Fig. 8.13 Trim actuator.

actuator consists of a barrel nut, rotated by a small sprocket, and a multi-start actuator screw threaded into the nut. Rotation of the nut by the control chain moves the screw in or out as required. Inspection centres primarily on the combination of screw and nut for excessive wear. The actuator screw should be removed from the actuator body and degreased. As much grease as possible should be removed from the barrel nut. The screw is mated with the nut by one or two turns and the amount of end play measured. The procedure is repeated at intervals until the screw is fully in. Where tolerances are quoted in the manual there is no problem in assessing wear but where this information is not available an assessment will have to be made by comparison with a new unit. In any case the total free play in a trim tab when measured at the trailing edge should not exceed 2½% of the maximum chord and this allowance includes any free play in the actuator. Check that the bearing on which the nut rotates is seviceable and turns freely and smoothly. The screw is hand-packed with grease on re-assembly. Larger screw jacks may be used with variable incidence tailplanes but the inspection is similar.

Control system rigging

The rigging of any control system follows the same basic principles although the physical layout may vary greatly. The basic consideration is the location of both ends of the system in neutral followed by connecting the two ends together. This may be complicated further by intermediate components which must also be in their neutral position. To achieve this one or two basic tools are necessary.

In order to measure cable tension an instrument known as a cable tensiometer is used (fig.8.14). In use it is simply clipped on to the cable in the middle of a long, clear

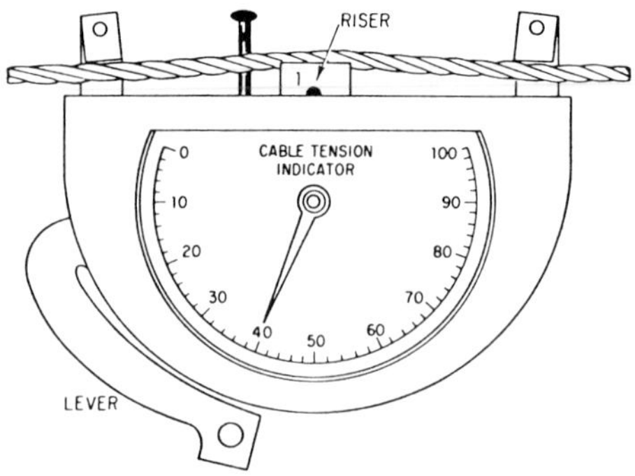

Fig. 8.14 Cable tensiometer.

span and the reading on the dial noted. This reading is converted to cable tension by means of a calibration graph or table supplied with the instrument. The instrument may be used on cables of different sizes by changing the centre riser. The calibration of a tensiometer may be simply checked by suspending a piece of control cable with a known weight on the end. The tensiometer is clipped on the cable and the reading compared with that in the calibration chart.

The range of movement of control surfaces is often expressed in degrees and may

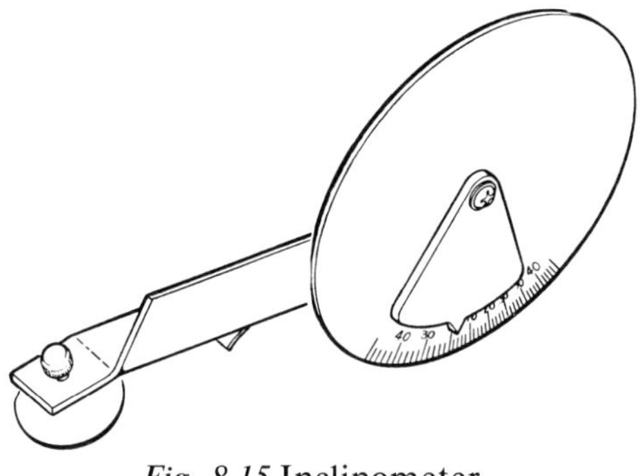

Fig. 8.15 Inclinometer.

be readily measured with a simple inclinometer (fig.8.15). Whilst this is suitable for aeroplane control surfaces it is not suitable for a helicopter swashplate and special tooling may be necessary. In many cases the bubble protractor from a combination set may be used successfully but it may be necessary to level the aircraft first to establish a suitable datum. In other cases the range of movement is quoted as a linear dimension, usually measured at the trailing edge of a control surface. In the case of an angular movement of a rudder it may be possible to transfer the movement to the floor with a plumbline, or to a piece of card, on which the angle may be measured with a protractor.

Rigging procedures

To illustrate the principles of rigging the procedures for typical control systems will be described but the principles must be applied in a manner dictated by the layout of the particular aircraft system.

Rigging a primary control system

In the aileron system shown in fig.8.1 the first need is to locate the pilot's control wheels in the neutral position. This is achieved by inserting a rigging pin (or the control lock) through holes in the instrument panel collar and the control wheel shaft. If dual controls are fitted the co-pilot's wheel must be brought into line by rigging the interconnecting chain system and checking the alignment of the control wheels with a straightedge. The cables are now correctly routed through the aircraft and loosely connected, making sure at this stage that the ailerons operate in the correct sense. The cables are now evenly tightened using the turnbuckles, making sure that the two aileron bellcranks are in the centre of their range of movement. (Sometimes the neutral position of the bellcrank is checked with a rigging pin but in other cases it is by measurement.) The cable tension should be checked on the balance cable and the turnbuckles locked when this is correct.

The push-pull rods are now adjusted to bring the ailerons to their neutral position which, in this particular case, is when the trailing edges are in line with the trailing edges of the flaps. Since both threads on the push-pull rods are right hand it will be necessary to disconnect one end of the rod to make the adjustment. Once the adjustment is correct lock the rod ends with the lock nuts and check that they are in safety. The rigging pin in the control column may now be removed and the controls checked for smooth and free movement. Pay particular attention to the push-pull rods to ensure that they are not binding at the extremes of travel. Check the sense of the movement to make sure that the ailerons move in the correct direction.

With an inclinometer mounted on one aileron check the range of movement and adjust the bellcrank stops as necessary. (If additional stops are fitted to the control columns ensure that a small clearance exists at each stop at the extremes of travel.) Now mount the inclinometer on the other aileron and check that the travel is correct. Record the actual travels and tensions on the worksheets for the job. Attach a small spring balance tangential to the outer edge of the control wheel and measure the force required to rotate the control. This is known as a static friction check and should not exceed 2 lb (0.9 kg) for an aileron system. The comparable values for other systems are: elevator - 4 lb (1.8 kg), rudder - 6 lb (2.7 kg).

Check round the system making sure that cables are correctly routed, all parts are in safety and properly locked, all pulley guards are in place and that all rigging pins, tools and other materials have been removed from the aircraft. A duplicate inspection of the system should now follow and the content of this will be described later.

Rigging a trim tab system

The elevator trim system shown in fig.8.7 will serve as our example for the rigging of a typical trim system. Most of the features of a typical system are embodied, although different layouts may be used depending upon the control surface being trimmed and the aircraft manufacturer's preferences. As a general principle the main control system should be rigged first and then the control surface locked in neutral. Where a variable incidence tailplane is fitted it must be set in the proper position or the range of movement may not be correct in either the elevator or the elevator tab. In the tab system used as our example the principle of having both ends of the system in neutral is followed as before in that both chains must be equally disposed around the sprockets. (At either extreme of the travel the chain from the control wheel must have at least two links clear of the sprocket in order to provide a certain amount of cushion in the control.) With the chain at the tab actuator meeting the same conditions the cables may be correctly routed and connected. The cable tension may now be adjusted to the correct value and the turnbuckle checked for safety and locked. With the elevator tab push-pull rod disconnected from the actuator move the tab a specified 'up' position beyond the normal range, measuring the angle by using an inclinometer. The screw of the actuator should now be unscrewed until it is possible to connect the push-pull rod to the actuator without moving the tab from its predetermined position. In this system the stops are mounted on the trim cables in the aft fuselage. Locate the aft block against the cable terminal and tighten the bolt. With the other two blocks loose on the cables rotate the trim wheel until the tab is in the correct 'up' position, as measured by the inclinometer. Slide the centre stop block along to touch the aft block and clamp in position. Rotate the trim wheel until the trim tab is in its correct 'down' position and slide the forward stop against the centre stop and tighten in place. As these movements and adjustments are being made the sense of the system should be checked.

The trim tab position indicator consists of a spring-loaded pointer, one end of which runs in a spiral groove in the control wheel. With the tab in the neutral position, that is streamlined with the elevator, the indicator should read 0°. The elevator trim wheel may be removed from the flange on its shaft in order to permit alignment of the indicator with the remainder of the system. As before checks should be made that the cables are correctly routed, all parts are in safety and properly locked, all pulley and sprocket guards are in place and all tools and other materials have been removed from the aircraft. A duplicate inspection should follow. As mentioned before the maximum free movement at the trailing edge of the tab should not exceed 2.5% of the maximum chord of the tab if the mechanism is to be considered serviceable. Tabs are not usually balanced and excessive free play will result in tab flutter developing and causing control difficulties and even more rapid wear on the components of the tab control system.

Some light aircraft do not use a control surface tab for trim but rely on an adjustable spring to trim out the control forces. This is particularly true of the rudder system although some older design aircraft use the same method for the elevator. Rigging usually involves putting the control surface (and also the nosewheel in the case of a rudder system) in the neutral position, ensuring that the trim wheel and indicator are also in the neutral position.

Power operated flap systems

Manually operated flap systems are rigged in a similar manner to any other manually operated flying control system, and for that reason only powered systems will be considered in this book. The power may be provided either hydraulically or electrically but the latter now predominates in the field of single and light twin-engined aircraft. Hydraulic actuation will be considered in Chapter 9 whilst electric actuation will now be considered.

Rigging an electrically powered flap system

There are many variations in design of electrically powered flap systems including those with preselected flap travel, but only the basic system will be considered here.

In the system illustrated in fig.8.16 the right wing flap and the motor must be correctly rigged before it is possible to correctly rig the left flap. Prior to installation

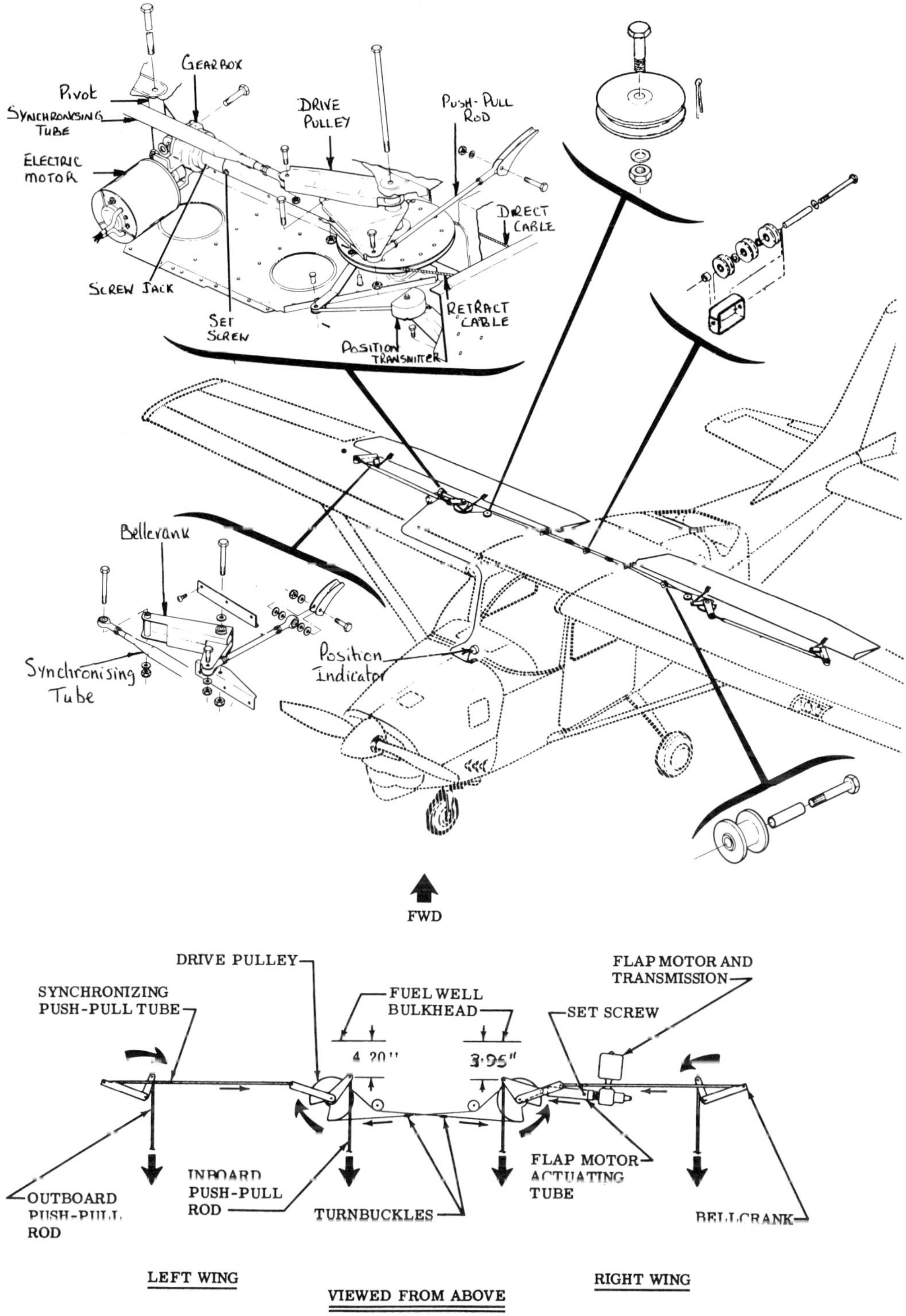

Fig. 8.16 Electric flap system.

the synchronising push-pull tubes in the left and right wings must be set to a specific dimension and then fitted to their bellcranks and drive pulleys. The motor is now carefully run to the position where the screwjack is fully retracted and the set screw is slackened. The drive pulley is positioned as shown and the screwjack assembly of the actuator rotated until it mates with the drive pulley. The set screw is tightened and the actuator connected to the drive pulley and the dimension rechecked. The right flap is now raised manually and the push-pull rods adjusted to mate with the flap in the full-up position. Due to limited accessibility the rods are connected first to the flaps and then to the bellcranks.

The cables are now connected at the turnbuckles and adjusted to locate the left drive pulley in the position shown whilst maintaining the correct cable tension. The left flap push-pull rods may now be adjusted to match the bellcranks in the same manner as for the right flaps. Operate the system and check that the cable tension is correct at intermediate positions and lock the turnbuckles. Recheck drive pulley rigging dimensions and all lockings. Finally check the flap travel using an inclinometer and synchronise the position indicator with the actual flap position.

The system described uses an overunning clutch to prevent damage to the motor at the end of the travel and the stops are mechanical. Other systems use microswitches at each end of the travel to switch off the motor and these must be adjusted to obtain the correct range of movement.

Helicopter control systems

At first glance the helicopter control system appears to have little in common with its counterparts in the field of the aeroplane (fig.8.17). Yet a closer study will reveal many similarities. The standard hardware is the same in that control rods and tubes, bellcranks, cables and pulleys all feature as the means of transmitting the motion from the pilot's controls to the control surfaces, in this case the rotors. The basic principles used in rigging the control system in a helicopter are the same as those used in rigging any other control system. However there are basic differences. As an example the cyclic stick, controlling both lateral and longitudinal motion of the helicopter, controls only one rotor, unlike the separate aileron and elevator controls in the fixed-wing aircraft. The cyclic control system accomplishes this double function through use of a mixer unit which takes different forms according to the helicopter type. The collective control may also be connected to the mixer unit. In the system illustrated the vertical cyclic control rods both move in the same direction for fore and aft movement of the cyclic stick but move in opposite directions when the stick is moved laterally.

Interposed between the stationary parts of the control system and those that rotate with the main rotor is the swashplate (fig.8.18). There are many variations of design but all consist of two parts, an upper, rotating part and a lower, fixed part. The upper part rotates with the mast and the main rotor head and is driven by them through a set of torque links. Equally spaced horns and rods connect the rotating swashplate to each blade. The lower swashplate is stationary and is prevented from rotating by a second set of torque links but receives cyclic and collective inputs from the pilot's controls. The whole swashplate is free to move vertically to obtain collective control of the rotor and is free to tilt in any direction to obtain cyclic control. This is achieved by mounting the lower swashplate on a spherical bearing which is able to slide vertically on the swashplate support. Thus in collective control the entire swashplate moves vertically but in cyclic control it tilts about the spherical bearing. The bearing

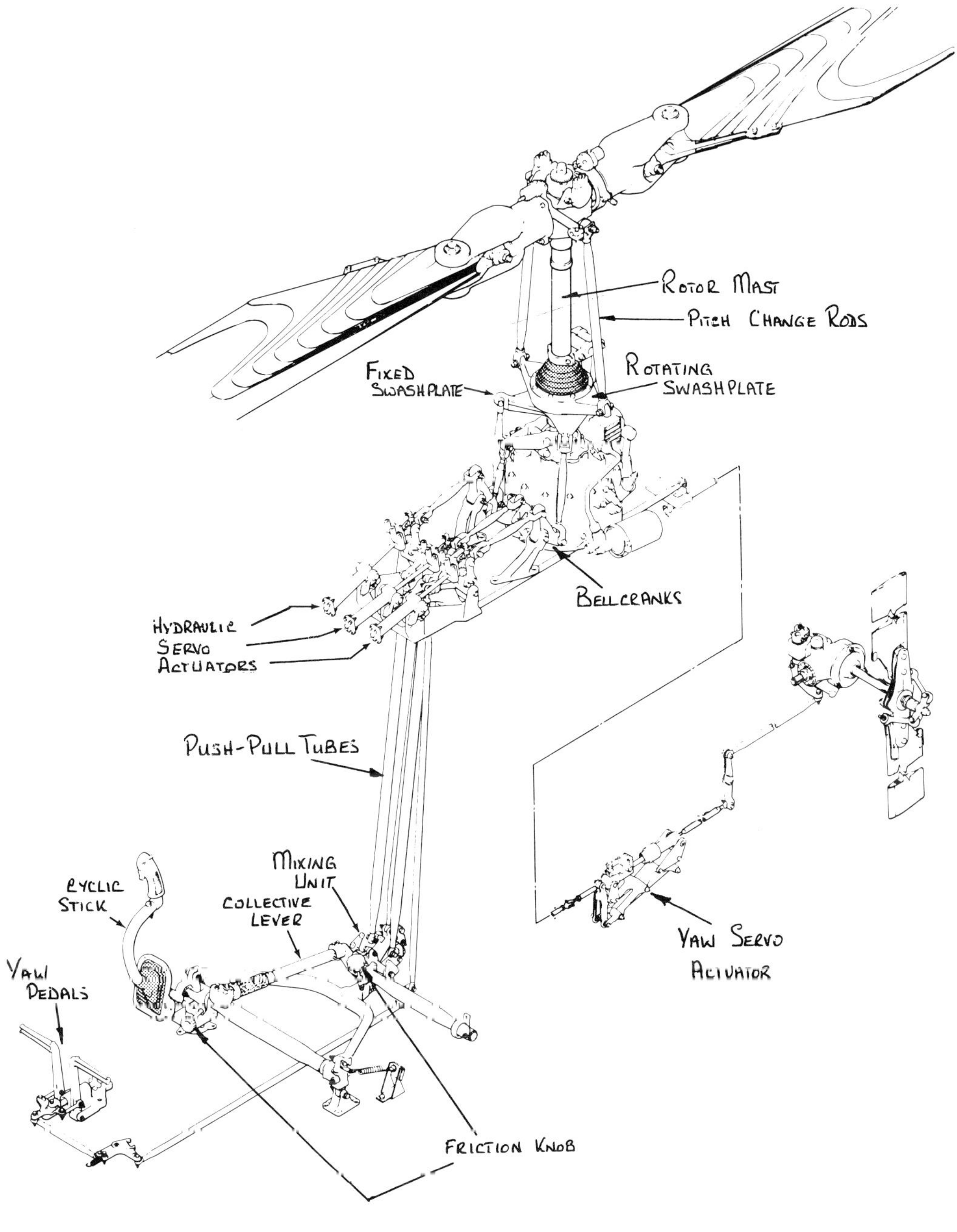

Fig. 8.17 Helicopter control layout.

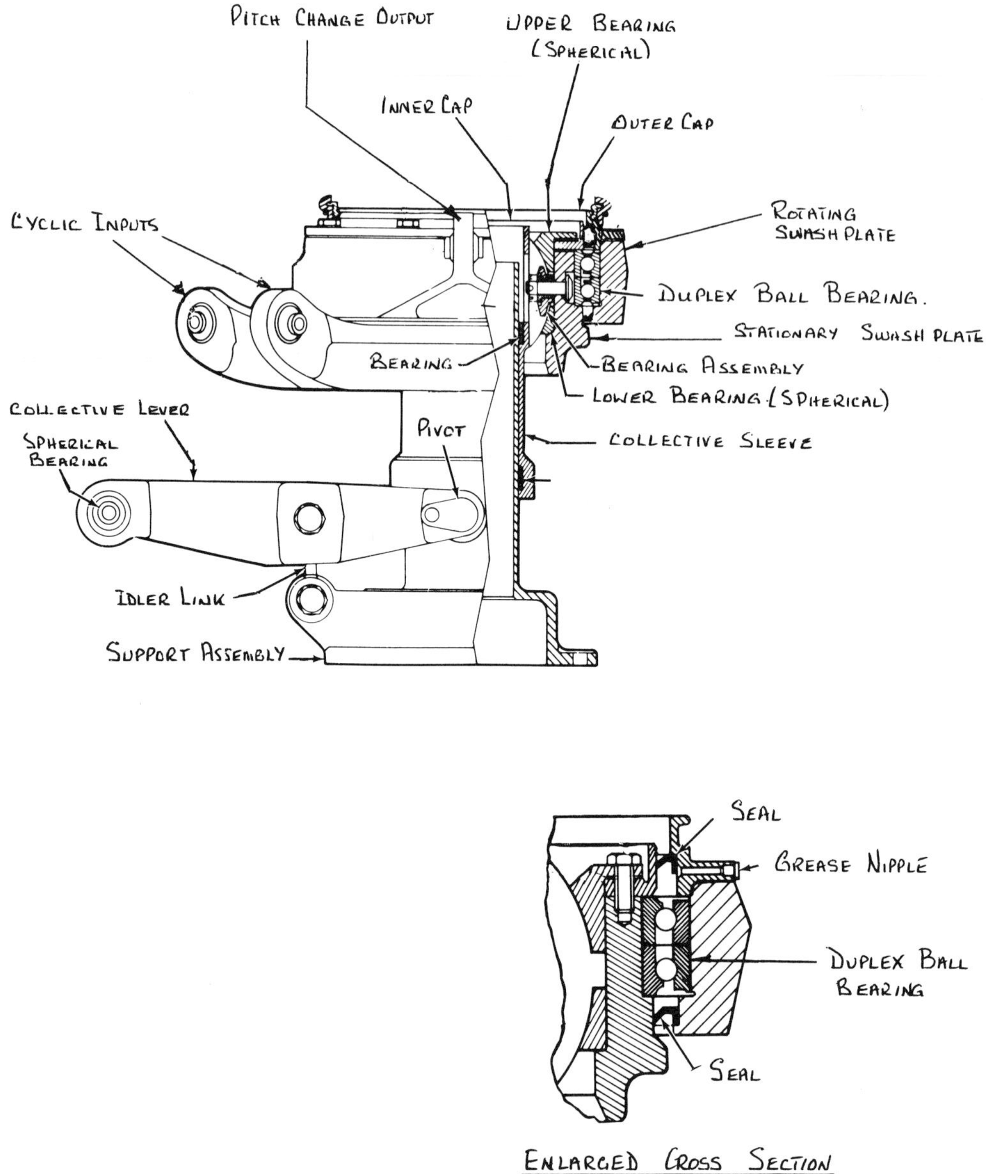

Fig. 8.18 Swashplate assembly.

in the swashplate is usually grease lubricated whereas the spherical bearing often incorporates a liner of synthetic material such as teflon that does not require additional lubrication. Wear in the swashplate and control rod bearings will cause vibration in the main rotor system and for this reason these items must be checked periodically. There must be a certain degree of stiffness in the movement of the swashplate on the spherical bearing and this should be periodically checked by means of a pull-off check using a spring balance. It is adjusted by changing the thickness of the shims between the two halves of the swashplate assembly to give greater or lesser 'nip' on the spherical bearing.

Rigging the swashplate

Whilst it will be necessary to make fine adjustments to the control rigging in order to have a smooth helicopter the heart of the rigging is the basic setting of the swashplate. In the system illustrated the collective control is rigged first followed by the cyclic. A rigging pin is inserted in the base of the cyclic stick and the friction knob is tightened to lock it in the neutral position. The collective lever is set in a predetermined position and locked there with the friction knob, and the intermediate bellcranks and push-pull tubes may now be adjusted and connected. The collective lever is positioned against the low pitch stop and the upper push-pull tube is adjusted and connected to the collective sleeve lever in such a way as to position the centre of the sleeve connecting pin a known distance above the top surface of the gearbox.

With the collective control rigged as far as the swashplate the cyclic controls may be rigged. With the rigging pin still engaged in the cyclic stick the intermediate push-pull tubes are adjusted to position the intermediate bellcranks in a known position. This is found by measuring from the centre of the tube attachment bolt to the deck on which the bellcrank is mounted. The swashplate is levelled by ensuring that measurements from each control horn to the top of the gearbox are equal and the final push-pull tubes adjusted to fit and connected up. The rigging of the pitch change rods to the main rotor will be dealt with in another chapter.

Rigging the tail rotor controls

The principle employed is again that of having both ends of the system in known, related positions and connecting up the intermediate linkage. In the case of the system illustrated (fig.8.19) the crosshead on the tail rotor is pushed fully in and the

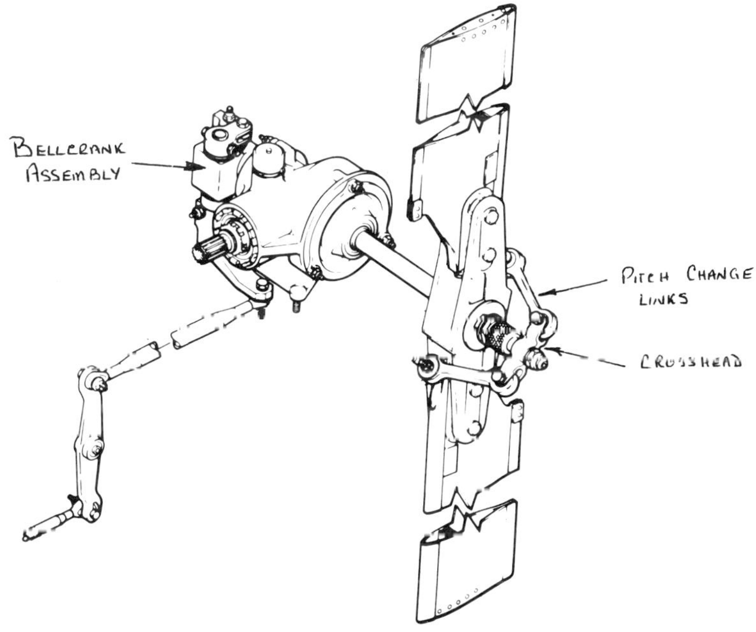

Fig. 8.19 Tail rotor control.

left pedal is pushed fully forward. The intermediate bellcranks are positioned and the push-pull tubes adjusted and connected. In some helicopters the collective and tail rotor controls are interconnected so that with increasing collective pitch, and torque, there is automatic adjustment of the torque reaction provided by the tail rotor.

Post adjustment inspections

Whatever the system that has been rigged there are a number of important inspections that must be made on completion of the adjustments and checks. In the U.K. there is a requirement for these inspections to be carried out by a second qualified engineer and they are embraced by the term duplicate inspection. This is an inspection that has been carried out by a qualified engineer and is carried out a second time, or duplicated, by a second qualified engineer. A duplicate inspection is required after all work affecting any control system that is used to alter the altitude of the aircraft or change the propulsive force of the engines. This means that all the primary and auxiliary control and trim systems are included as are slots, slats, speedbrakes and the like. The attachment of rotor blades in a helicopter is also the subject of a duplicate inspection.

The depth of the inspection will depend upon the nature of the work carried out, which could vary from a simple adjustment to complete overhaul of the system. Essentially it includes all the control components for correct routeing, safety, locking, proper pulley and chain guards and correct assembly. It also includes the operation of the controls in the correct sense over the correct range of movement with the correct friction and without fouling anything else. Finally the duplicate inspection would include a check that all tools, rigging pins, rags and other materials have been removed from the aircraft. The inspection is complete when both engineers have made the appropriate certifications in the aircraft records.

CHAPTER 9
Hydraulic Systems

General considerations

A basic hydraulic system consists of relatively few components although their size and complexity may vary considerably with the specific type of aircraft to which they are fitted. At the very lowest level a workable system would include a reservoir for the working fluid, a hand pump to provide the pressure, a selector valve to control the operation of the circuit and an actuator cylinder to provide the movement (fig.9.1).

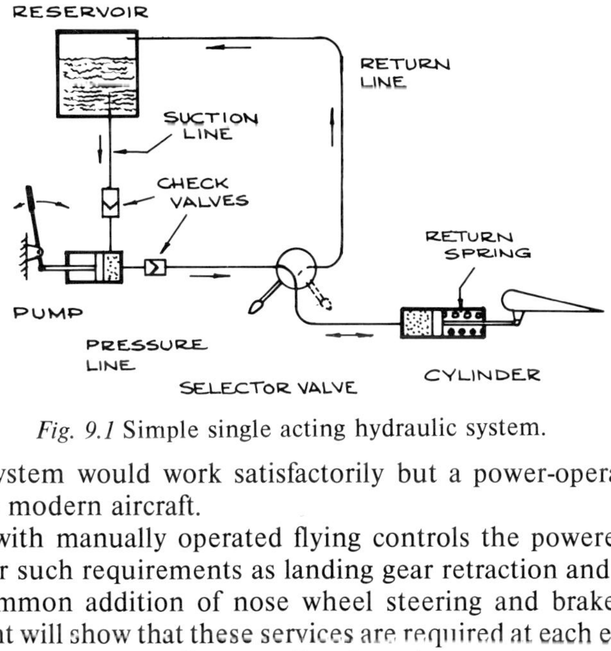

Fig. 9.1 Simple single acting hydraulic system.

Such a simple system would work satisfactorily but a power-operated system is necessary for the modern aircraft.

In an aircraft with manually operated flying controls the powered services are primarily used for such requirements as landing gear retraction and flap operation with the less common addition of nose wheel steering and brake operation. A moment's thought will show that these services are required at each end of the flight but not normally between take-off and landing. In order to reduce wear on the pump, and also the temperature of the hydraulic fluid, one of two methods is adopted to relieve the pump of pressure when it is not required. In the first method a pressure-operated unloader valve will allow the fluid from the pump to circulate back to the reservoir without restriction any time the main system pressure is up to the required value (fig.9.2). A drop in the system pressure will cause the unloader valve to route the fluid under pressure into the main system again. The second method utilises a special design of selector valve which permits the fluid to flow freely through the valve and back to the reservoir as long as all the valves are in their neutral position (fig.9.3). The first alternative is known as a closed centre system and the latter as an open centre system.

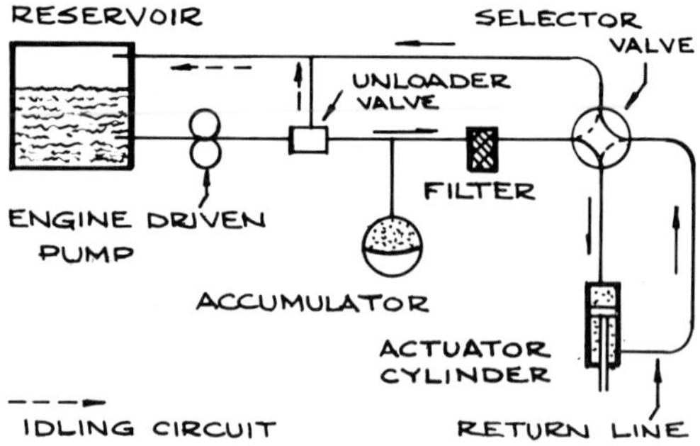

Fig. 9.2 Closed centre hydraulic system.

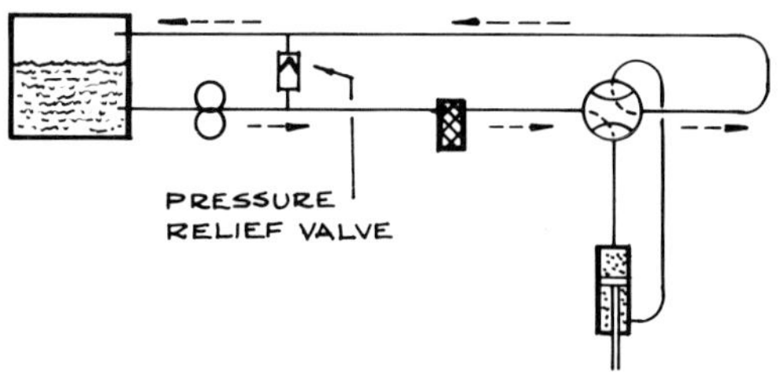

Fig. 9.3 Open centre hydraulic system.

Operating pressures vary considerably with the type of system that is employed and the forces that are required to be developed by the actuators. Hence a simple power assistance system for helicopter flying controls may have an operating pressure of only 750 p.s.i. (52 bars) whereas a system that must operate flaps and landing gear in a typical executive aircraft will have a working pressure of 1 200-1 500 p.s.i. (83-104 bars). Just one or two aircraft in the General Aviation bracket have working pressures of 3 000 p.s.i. (207 bars).

Hydraulic fluids

The hydraulic fluid predominantly used in the class of aircraft being considered is a mineral-based fluid produced to meet one of the national aeronautical specifications. In the U.S.A. this specification is MIL-H-5606 and in the U.K. it is DTD 585. The two specifications are compatible with each other and the fluid is dyed red for identification purposes. However, a fluid with increased fire resistance based on a phosphate ester has been developed and is commonly known as Skydrol. This synthetic fluid is used in a couple of General Aviation aircraft and is definitely not compatible with the mineral-based fluid. It is susceptible to contamination by atmospheric moisture, will destroy the seals used with mineral-based fluids and strip all paints other than epoxy or polyurethane based finishes. The engineer should never handle it unless wearing goggles and gloves for some fairly obvious reasons. However, protective clothing must be suitable for the job since Skydrol will attack PVC, a material commonly used for protective gloves. Quite apart from these precautions absolute cleanliness is a must for satisfactory servicing of a hydraulic system.

Dirt in the system, however fine, will be carried throughout the system by the fluid and will abrade the surfaces of moving parts that are made to fine tolerances. Always keep the reservoir tightly closed and use proper fitting caps to seal the lines and fittings at any time a component is removed. To check the fluid for contamination drain a sample from the system into a glass laboratory beaker and view against a normal tungsten bench lamp. A sharp twist of the beaker will reveal solids in suspension as they swirl in the fluid. Old and contaminated fluid will be darker in colour than fresh fluid and should be drained off. The system is then flushed by refilling it with fresh fluid, circulating it through the system and again draining. Finally refill with fresh fluid. Components must be cleaned with a solvent that is compatible with the fluid in use. Where a mineral-based fluid is in use Stoddard solvent (dry cleaning solvent) is suitable but only trichlorethylene should be used in a Skydrol system.

Drawings used by the various manufacturers may vary quite widely in the symbols used and the style of presentation. It is important to familiarise yourself thoroughly with the particular method of representation in order to have a proper understanding of the system operation. Just occasionally there are traps for the unwary; for example, the French appear to draw their non-return valves the opposite way round to normal British and American conventions!

Fluid system hardware

As in the case of flying control systems there is much hardware that is common to both large and small systems.

Rigid pipelines

Probably the least noticed parts of a hydraulic system are the rigid pipelines that connect the various components together. The reason for this is quite simply that we have become accustomed to a very high degree of reliability in the pipelines and therefore tend to take them for granted. However, proper manufacture, assembly and maintenance are very important if that level of reliability is to be maintained. At the pressures used in the hydraulic systems of light aircraft the predominant choice of material for rigid tubing is light alloy, usually a magnesium alloy in the annealed condition. The most commonly used specification in the U.S.A. is SAE 5052-0, the first 5 indicating a magnesium alloy and the -0 indicating the annealed condition. For high pressure hydraulic lines SAE 6061-T6 or even SAE 2024-T4 is used. In the U.K. the choice for a magnesium alloy tube falls with L.56. Whilst the materials mentioned have a high degree of corrosion resistance it is not unusual to find pin-point corrosion on magnesium alloy lines that have been in service for some years. This can often be readily found by feel, the surface of the tubing feeling rough to the touch. If left unattended the corrosion will progress through the wall of the tube until a hydraulic leak develops under pressure, spraying a fine mist of hydraulic oil over the surrounding structure and components. The early stages of such corrosion can be dealt with adequately by rubbing the affected pipelines with an oily rag but in its later stages replacement is the only treatment.

The most common method of attaching the ends of the tubing to the fittings on the components is by flaring (fig.9.4). In this the end of the tubing is expanded by a special tool into a cone which mates with a matching cone on the fitting. The flare has wide applications outside the aircraft industry but the cone angle is not always the same as that used for aircraft. The angle is specified as the angle between the flare and the axis of the tubing, that is half the included angle. In the automotive world this is

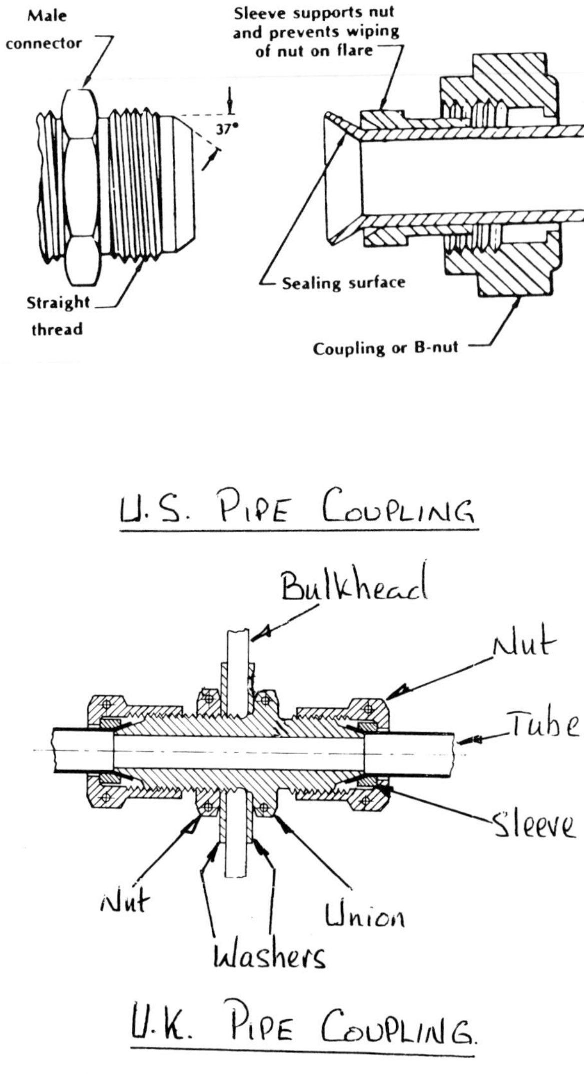

Fig. 9.4 American and British flared fittings.

often 45° but the U.K. aircraft industry uses an angle of only 16°. To confuse the situation still further the U.S. standard is 37°. The important point is that use of the wrong tool will prevent the flare sealing correctly.

The correct flaring of a tube end is not a difficult procedure but does require a certain amount of care or the final result will be unsatisfactory. The tube must first be cut perfectly square using a tubing cutter or a fine-tooth hacksaw. When using a tubing cutter the cutting blade is fed into the tube as it is rotated round it. After the tube has been cut remove the burrs from the inside edge of the tube with a small scraper. The end and outside diameter of the tube should be smoothed with a fine file or abrasive cloth, working round the end rather than across it. It is important to make sure that there are no chips or burrs on either the tube or the flaring cone since these will become embedded in the flare and weaken it. It is equally important to make sure that all nicks and scratches are polished out of the end of the tube or there is a good possibility that the flare will split when it is formed. Insert the end of the tube into the flaring tool to the correct distance and clamp it in position. Lubricate the cone of the tool and rotate the handle until the flare is the correct size. The outside diameter of the flare should extend approximately $\frac{1}{16}$ in. beyond the end of the sleeve, but should not be larger than the outside diameter of the sleeve. Before flaring the other end of the tube make sure that any nuts and sleeves have been correctly threaded on to the tube.

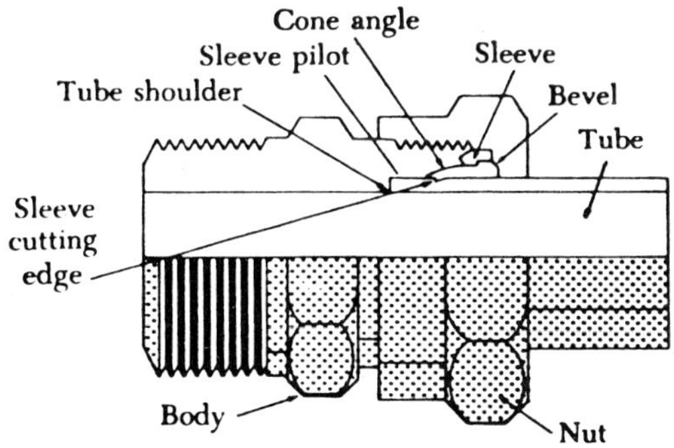

Fig. 9.5 Flareless fitting.

An alternative to the standard flared fitting is the flareless fitting (fig.9.5). The tube should be prepared in the same manner as if a flared end was required. The nut and special ferrule are slipped on to the tube, ensuring that they are the correct way round. The tube is then inserted into a presetting tool, or if this is not available a steel fitting of the correct size, until the tube contacts the internal shoulder. The ferrule should be lubricated and the nut screwed down finger-tight whilst the tube is kept tightly against the stop. From this position the nut should be tightened a further $1\frac{3}{4}$ turns to preset, or crimp, the ferrule on to the tube. If the crimping has been satisfactory there should be a uniform ridge of metal raised above the surface of the tube at least half the thickness of the front edge of the ferrule. The ferrule should be slightly bowed and be free to rotate although not free to move axially.

Whilst on the subject of rigid pipelines we must give some attention to the bending and installation of these components. No tube, regardless of its length, should be installed unless there is at least one bend in it. This will make provision for expansion and contraction and help to absorb vibration. Almost all tubing used in aircraft fluid systems is thin wall and special care must be exercised when bending it to avoid flattening or kinking the inside radius of the bend. A table of typical minimum bend radii is given in fig.9.6. The maximum amount of flattening that may be allowed is 25% of the original diameter of the tube. In order to achieve satisfactory bending of aircraft tubing one of the many types of small benders should be used. All these use a mandrel that is grooved in such a way as to support the sides of the tube whilst it is bent to the required angle with a lever or screw-operated device.

When installing a rigid line it should never be pulled into place by tightening the nuts on the fittings. To do this will place a strain on the flare or the preset bite causing the pipe to fail in subsequent service, especially when subject to vibration. The pipe must fit properly into place and it must be possible to screw the nuts down using only the fingers. The nut of a flareless fitting should be tightened a further $\frac{1}{6}$ to $\frac{1}{3}$ turn to complete the connection. In a flared-type fitting the nut should be tightened to the torque given in the table for American fittings. Torques are not specified for British fittings and tightening is a matter of experience. Should a fitting leak after assembly

Tube O.D. (inches)	Torque range for tightening tube nuts (lb.-in.)		Minimum bend radii (inches)	
	Alum. alloy 1 100-H14, 5052-O	Steel	Alum. alloy 1 100-H14, 5052-O	Steel
1/8			3/8	
3/16		30–70	7/16	21/32
1/4	40–65	50–90	9/16	7/8
5/16	60–80	70-120	3/4	1 1/8
3/8	75–125	90–150	15/16	1 5/16
1/2	150–250	155–250	1 1/4	1 3/4
5/8	200–350	300–400	1 1/2	2 3/16
3/4	300–500	430–575	1 3/4	2 5/8
1	500–700	550–750	3	3 1/2
1 1/4	600–900		3 3/4	4 3/8
1 1/2	600–900		5	5 1/4
1 3/4			7	6 1/8
2			8	7

Fig. 9.6 Bend radii and nut torques.

it should be dismantled in order to find the cause of the problem rather than just applying more torque. In addition to proper installation of the fittings the line should be supported in cushion clamps at intervals not exceeding 12 in. for a 1/4 inch line. This may be extended to 16 in. for a 3/8 inch line. Where a line passes through a bulkhead and a bulkhead fitting is not employed it must be supported so that it cannot touch the edges of the hole. In addition to using a clamp the sharp edge of a hole should be protected with a grommet.

Flexible hoses

Where there is relative movement in a system a rigid line cannot be used and a flexible hose is the answer. The high pressure flexible hose consists of a hose section that is reinforced with steel wire braid with a liner of synthetic rubber or teflon (tetrafluorethylene). Teflon has particularly good characteristics when used in fuel, oil, hydraulic or pneumatic systems, but there is one characteristic of which the engineer must be aware if service life is to be satisfactory. Teflon will take a set, that is form to a shape in which it is restrained, after it has been used with high temperature or pressure fluids. When it is necessary to remove such a hose from the aircraft temporarily the ends should be joined together with locking wire or string in order to keep it in the shape that it had when installed. The hose portion may be bought in bulk and made up into a hose assembly with reusable end fittings. The procedure varies slightly with the type of hose that is being used but generally follows the procedure outlined here. For a specific type of hose always follow the manufacturer's instructions implicitly.

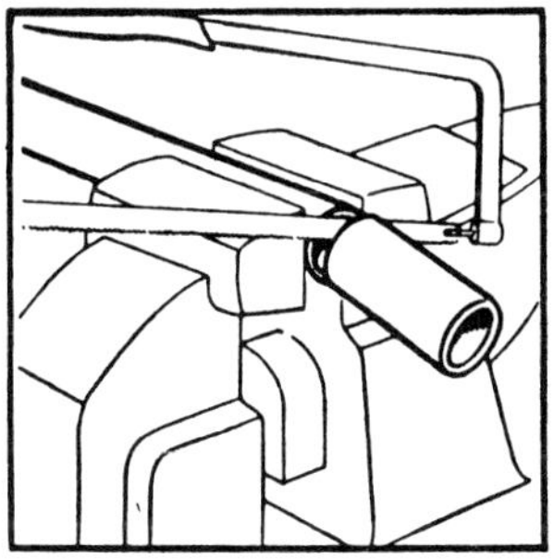

1. Place hose in vise and cut to desired length using fine tooth hacksaw or cut off wheel.

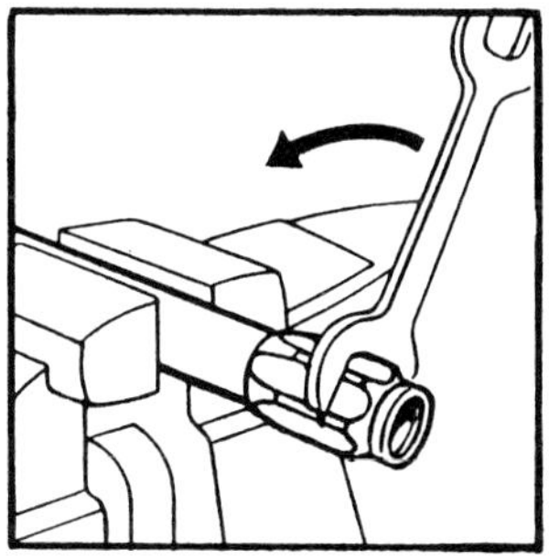

2. Place hose in vise and screw socket on hose counterclockwise.

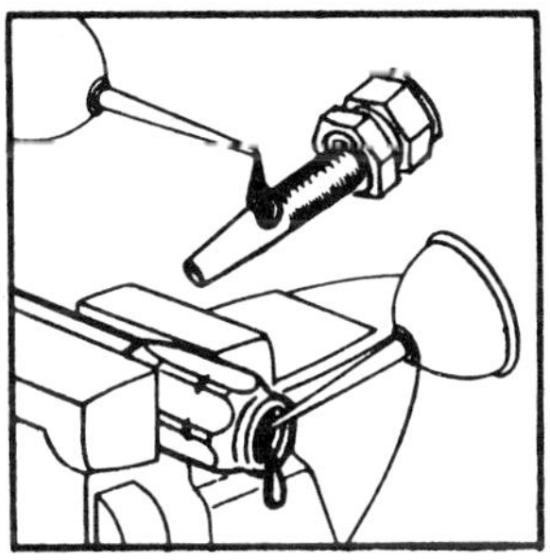

3. *Lubricate inside of hose and nipple threads liberally.

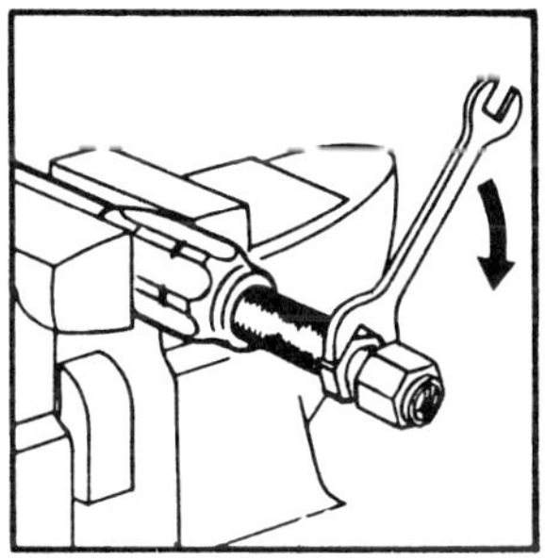

4. Screw nipple into socket using wrench on hex of nipple and leave .005 inches to .031 inches clearance between nipple hex and socket.

Fig. 9.7 Flexible hose assembly.

The assembly of flexible hoses will, of course, vary with the manufacturer but the instructions that follow are typical (fig.9.7). In order to facilitate cutting the hose squarely to length it should be wrapped with masking tape and the cut made through the tape with a fine hacksaw. This will prevent the wire braid flaring out as the cut is made. The tape is removed and the hose inserted into the socket with a twisting motion until the end is in line with the back of the socket threads. Mark the hose at the rear of the socket with a chinagraph pencil or with a piece of masking tape. The inside of the hose and the nipple threads should now be lubricated with an SAE 30 oil and the nipple inserted into the socket. Carefully engage the nipple and socket threads whilst holding the hose in position. Check the mark previously made to ensure that the hose has not been moved out of position. Complete the assembly by screwing the nipple down with a spanner until there is a specified gap between the nipple and the socket. This gap is dependent on the size of the hose. Make a final check that the hose has not moved in the socket, clean the assembly and pressure test.

Flexible hoses are commonly provided with a printed stripe, or manufacturer's identification, throughout the length of the hose. When the hose is installed it is important that this line is not twisted since a twist in the hose will cause a stress to be set up in the assembly which is likely to shorten its service life. When pressure is applied to a flexible hose there is a tendency for it to expand in diameter and to shorten in length and for this reason there should always be 2–4% slack in a hose assembly. The radius over which the hose is bent is important to maximum service life, the minimum radius being related to the diameter of the hose, and in addition there should be no bends near the end fittings of the assembly.

Pressure testing

Sometimes known as proof testing, pressure testing is always applied to any newly assembled flexible hose or rigid line in order to verify its integrity. The test pressure used is 150% of the normal maximum working pressure of the system in which the hose or line is to be used. Hence if a hose assembly is to be used in a hydraulic system working at 1 500 p.s.i. it should be pressure tested to 2 250 p.s.i. The method of pressure testing is simple and should be accomplished using the normal system fluid. If, however, this would result in excessive risk high pressure air or water may be used. To pressure test a hose or a pipe the end should be blanked with a pressure blank designed for the fitting in use. The hose is then filled with the test fluid, connected to the pressure source, and the pressure slowly raised to the test pressure. The pressure should be held at this value briefly and then slowly reduced to zero. During the test the hose should be carefully watched for any signs of leakage or abnormal distortion. The operator must be protected during the test by a proper enclosure with a window that is capable of resisting the full force of a burst hose. A simple high-pressure source for the small workshop may be made by adapting an automative diesel injector tester.

Fittings

The variety in pipe fittings is bewildering, with each national standard already mentioned having its own range of sizes and styles. The American standards differentiate between pipe and tube fittings on the basis that a pipe fitting incorporates a tapered National Pipe Thread on at least one end of the fitting. A tube fitting is one that is designed for flared or flareless fittings on rigid pipes (tubes). A word of caution. Early U.S. aircraft used flared fittings in the AC range and these are not compatible with the later AN and MS ranges. Fig.9.8 makes the differences clear.

One fitting and its installation deserves special mention and this is the American bulkhead fitting (fig.9.9). The bulkhead fitting, which may be one of many different styles, is identified easily in that one arm is longer and has two threaded portions separated by a land or groove. When this is installed through a bulkhead, or other structural member, a washer is used under the shoulder of the fitting and a second washer under the nut. These washers are of the same material as the fitting and the second one may be omitted if the bulkhead is over $\frac{3}{16}$ in. (4.8 mm) thick. Where this type of fitting is used in a casting, such as a brake cylinder, the installation technique is quite different (fig.9.10). First screw a nut on to the threads nearest the shoulder with the counterbore away from the shoulder. The upper threads of the fitting must not protrude into the counterbore of the nut. Then work a teflon backup ring over the lower threads of the fitting and into the counterbore of the nut. Cover the threads

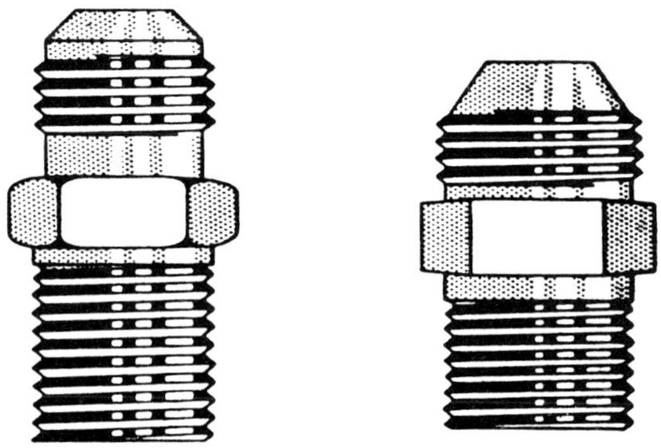

Fig. 9.8 AN and AC fittings.

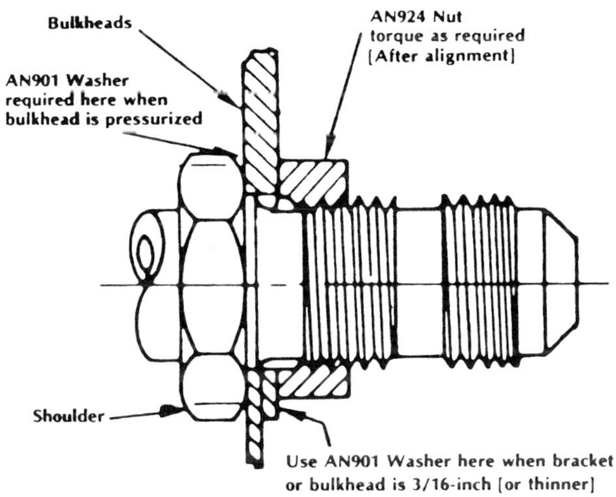

Fig. 9.9 Installation of bulkhead fitting.

with aluminium foil, paper tape or plastic, lubricate the O-ring with petroleum jelly and roll it carefully over the threads until it touches the backup ring. Screw the fitting into the housing until the O-ring contacts the chamfer, making sure that the nut and fitting are rotated together. Holding the nut stationary with a spanner the fitting should be screwed in 1½ turns plus a maximum of one additional turn to permit alignment with the hose or pipe to which it connects. The nut may now be tightened to the appropriate torque value whilst holding the fitting stationary with a spanner on the flats.

When installing a tapered pipe fitting into a component the first thread should be inserted into the hole and thread lubricant or sealant applied sparingly to the second thread. This will prevent galling of the threads without risk of contaminating the system. The fitting may now be tightened, taking care not to overtighten. When too high a torque is applied the tapered thread is capable of exerting sufficient force to crack the component into which it is screwed.

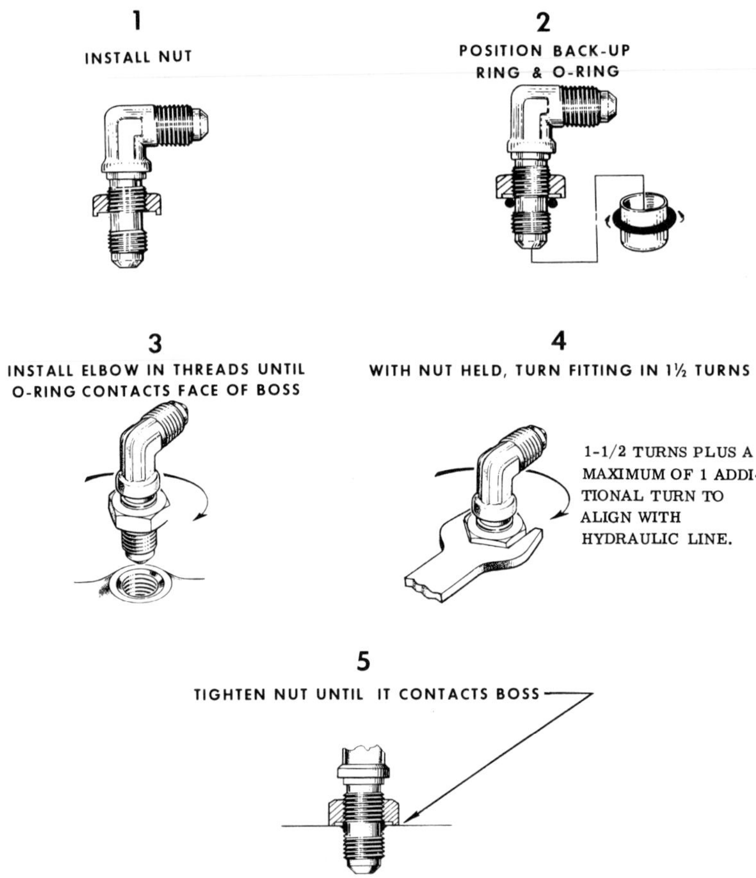

Fig. 9.10 Installation of fittings in components.

Seals

The most common type of seal used in hydraulic system components is the O-ring which because of its simplicity is often underrated. Perhaps there is no other component so small on which so much depends. For this reason great care must be taken in installing O-ring seals. Whenever an O-ring is to be installed over threads, flanges or other sharp edges always protect the ring by using aluminium foil, paper or plastic tape over the sharp corners. The seal should always be lubricated with the system fluid before installation. Never choose a replacement ring by appearance, always by part number. With different fluids in use O-rings are manufactured from different materials in order to be compatible with the system requirements and use of the wrong O-ring may result in failure. Where O-rings are installed in awkwardly located grooves hook and spoon shaped tools should be made locally from small diameter brass rod to facilitate installation and removal. These should be highly polished to prevent damage to the seals. Where an O-ring fitted to a piston, for example, is being inserted into the mating cylinder it is important that the edge of the cylinder is chamfered to a diameter greater than the outside diameter of the O-ring. Even so great care and adequate lubrication is essential to prevent damage to the ring.

In high pressure systems (above 1 500 p.s.i.), and some oleo-type landing gear struts, there is a tendency for the O-ring to extrude into the clearance between the moving parts. This can be avoided by the use of a spiral backup ring, often made from teflon. This is installed on the side of the O-ring opposite to the pressure or on both sides of the ring in the case of a double-acting cylinder. Before installation a check should be made that the ring is spiralled in the correct direction so that the scarfs are correctly located in relation to each other (fig.9.11).

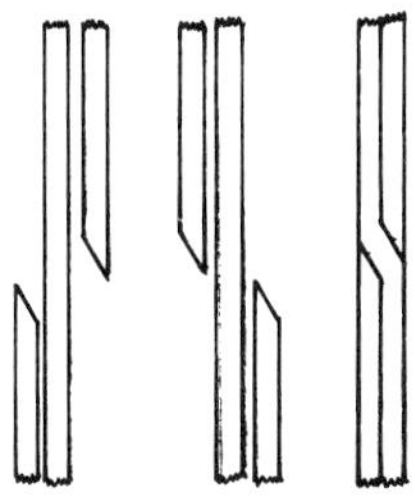

Fig. 9.11 Spiral backup ring.

Typical aircraft hydraulic systems

At the beginning of this chapter it was mentioned that there were basically two types of system, open centre and closed centre. A typical example of each of these systems will be described in order to study the functioning of the system and the components in the system. Although there may be substantial differences in the layout of the two systems there are also commonalities in such items as pumps, filters and actuators so some of these will be described first.

Since the majority of light aircraft are designed with relatively low pressure hydraulic systems a gear-type pump delivering about 1 gallon per minute is satisfactory (fig.9.12). The multiple piston type of pump is reserved for larger aircraft operating with much higher pressures and delivery rates. At one time the gear-type pump was always driven from one of the accessory drive pads on the engine but of late there has been an increasing trend towards the use of electrically driven pumps, and these will be discussed separately in this chapter. Where the pump is fitted to an engine accessory drive pad a drain line is installed to permit the free drainage of any hydraulic fluid that leaks past the pump seal or engine oil that passes the accessory drive seal. At all normal inspections the open end of the drain line should be inspected for any signs of fluid leakage and, if there is a significant amount, the source traced and corrected. The pressure and suction connections to the pump are made with flexible hoses, the smaller of the two being the pressure hose.

From the pump the hydraulic fluid is passed through a fine filter, often of the paper element type (fig.9.13). The element should be replaced at the intervals shown in the maintenance schedule, which is often at the 100 hour inspection. Before throwing the old element away cut it open with a sharp knife and inspect it for any signs of contamination that would reveal an impending failure of some component in the system.

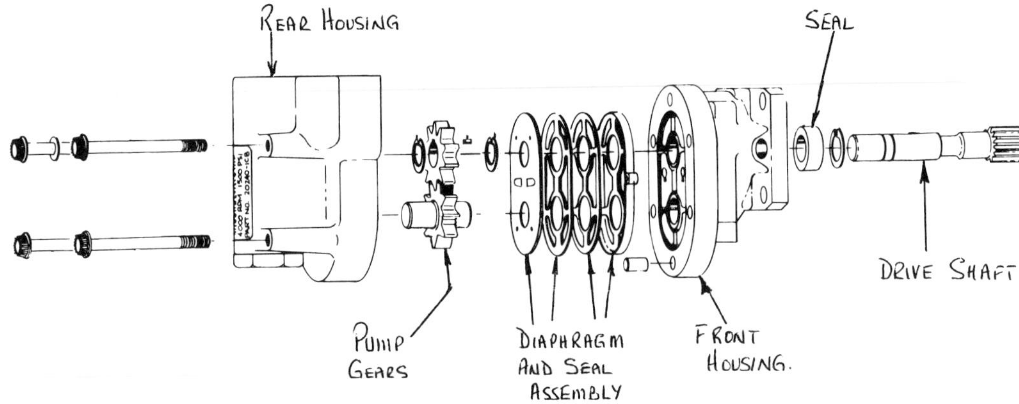

Fig. 9.12 Gear type pump.

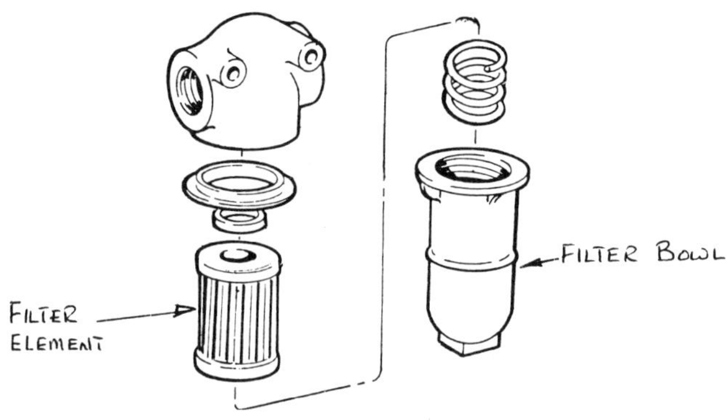

Fig. 9.13 Hydraulic filter.

At the other end of the system the hydraulic pressure is transformed into motion by the use of actuators which may be either linear or rotary (fig.9.14). Predominantly actuators are linear and are often known as cylinders or jacks. The size of the actuator will depend upon the size of the load to be moved and the hydraulic pressure available, being quite small in the case of a door actuator and considerably larger in the case of a main landing gear retraction jack. The rotary actuator used in some main landing gear systems is basically a linear jack in which the piston rod has been machined to form a rack. This rack engages with a pinion on the output shaft in order to convert the linear motion into rotary motion.

Open centre hydraulic system

Many systems used in light aircraft combine the fluid reservoir, the control and check valves and the emergency hand pump into one unit known as a powerpack. Such a unit is fed with hydraulic pressure from the pump, controls it in accordance with the pilot's demands and routes the pressure to the appropriate actuators. Such a system will be described using the sequence of diagrams in figs.9.15.–9.19.

With the selectors in neutral and the pump running the entire output of the pump is directed through the centre of each selector valve and back to the reservoir, hence

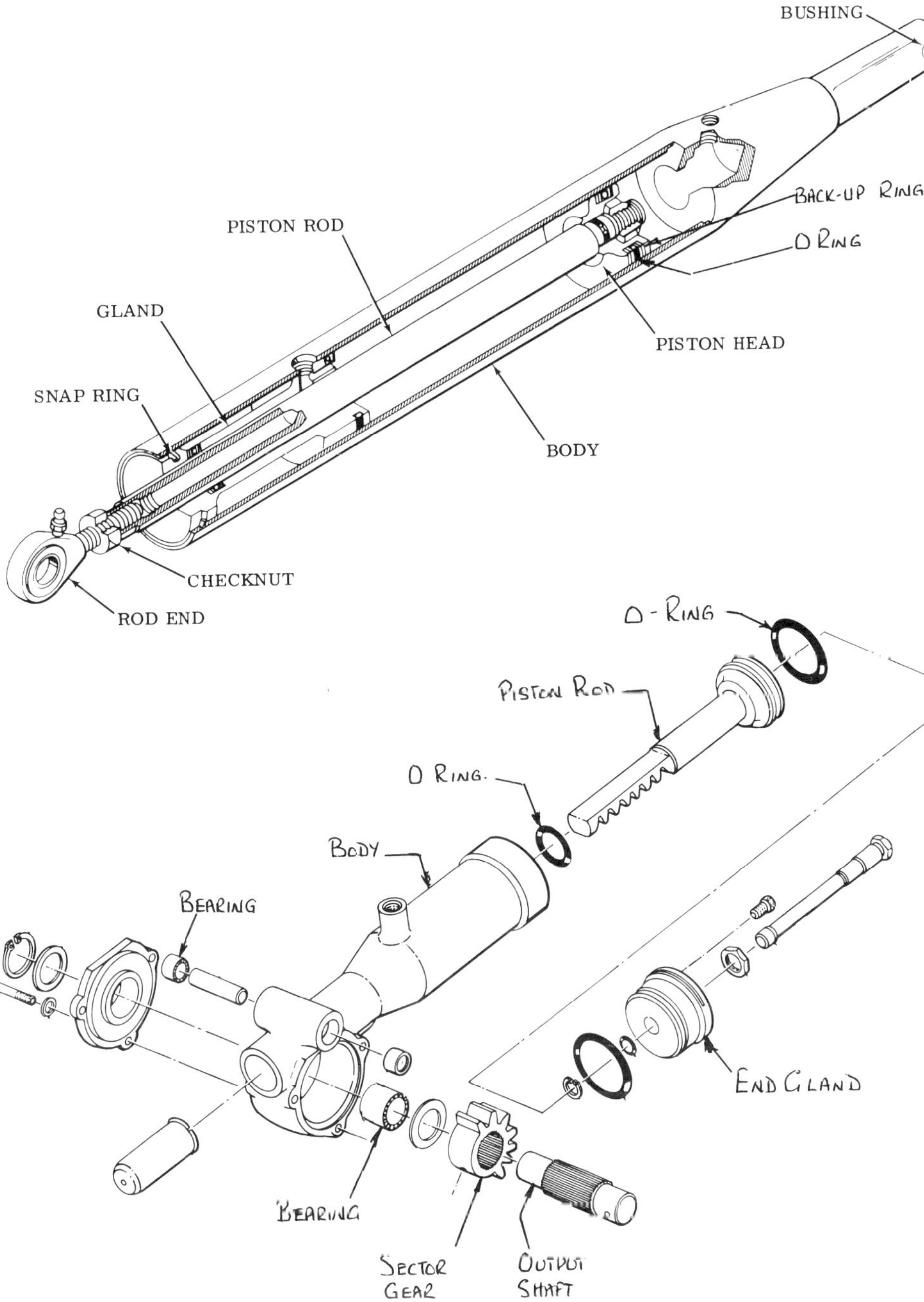

Fig. 9.14 Hydraulic actuators.

the name 'open centre' (fig.9.15). In this condition the pump produces just sufficient pressure to circulate the fluid and no more. Thus the pump is not loaded until such time as a pilot demand requires pressure for the operation of some system.

In fig.9.16 the landing gear has been selected up and the gear has moved almost to the fully retracted position. With the flow in the direction shown the priority valve

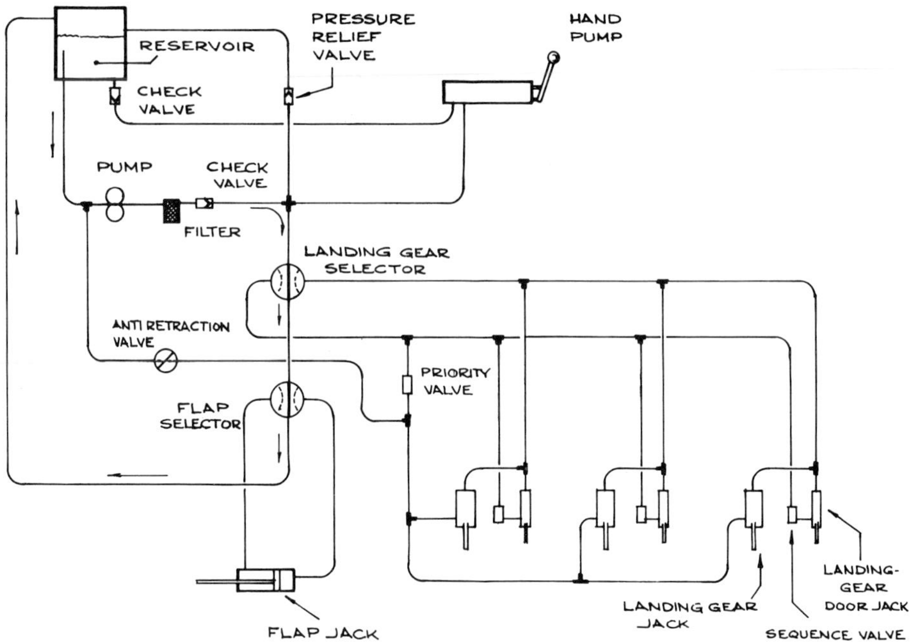

Fig. 9.15 Pump idling.

offers no resistance to the flow of the fluid. In fig.9.17 the landing gear has retracted fully, opened the sequence valves and the doors have closed. The sequence valves are in the system in order to prevent the doors closing before the gear has retracted. The pump, however, is still supplying pressure and this will build up until it is sufficient to release the selector lever detent and allow the lever to return to the

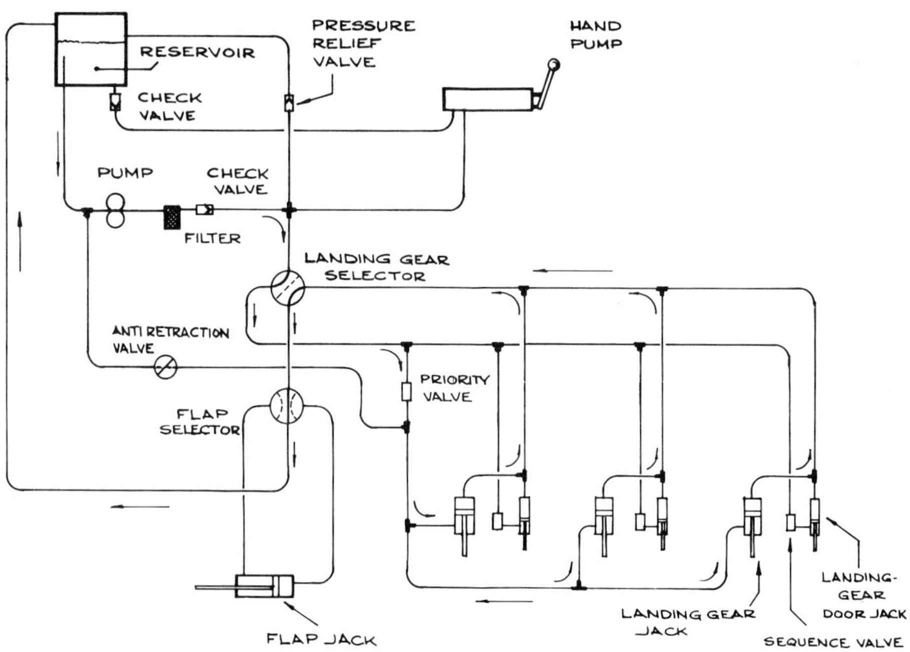

Fig. 9.16 Landing gear selected up.

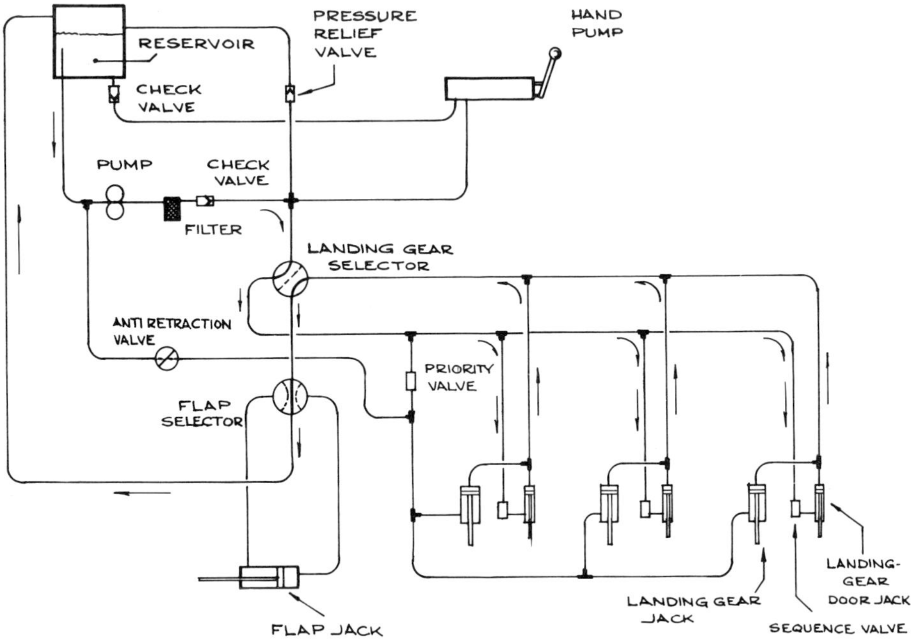

Fig. 9.17 Landing gear doors closed.

neutral position. The pump then idles through the open centre circuit.

Selection of the landing gear down (fig.9.18) results in the doors opening first since there is no restriction to the return flow of the fluid from the door actuators. The return flow of the fluid from the landing gear actuators is blocked by the priority valve which will not open until a pressure of 600 p.s.i. is reached. The actual figure will differ from aircraft to aircraft. Once the door actuators reach the end of their travel the

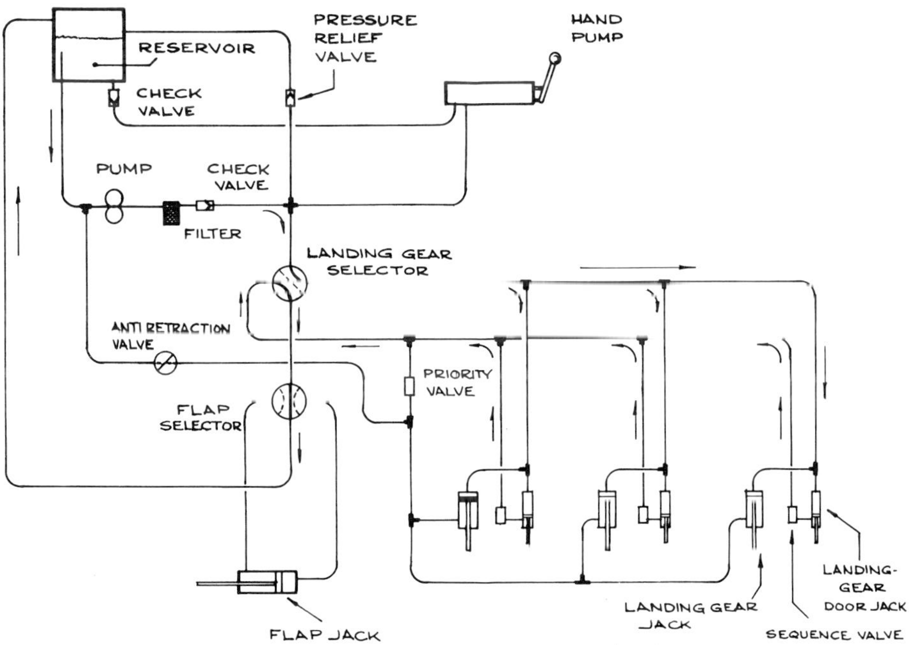

Fig. 9.18 Landing gear selected down.

system pressure builds up until the priority valve opens and allows the landing gear to extend (fig.9.19). The selector lever will be released back to the neutral postion when the system pressure reaches the detent value in the same way as already described.

The flap system is a simple system involving one jack and no other valves or components. It should be noted that the open centre arrangement with the landing gear selector upstream of the flap selector gives the priority to the landing gear. If the pilot selects the flaps down and then selects the landing gear down, the flaps will stop moving on the second selection because they have lost their source of hydraulic

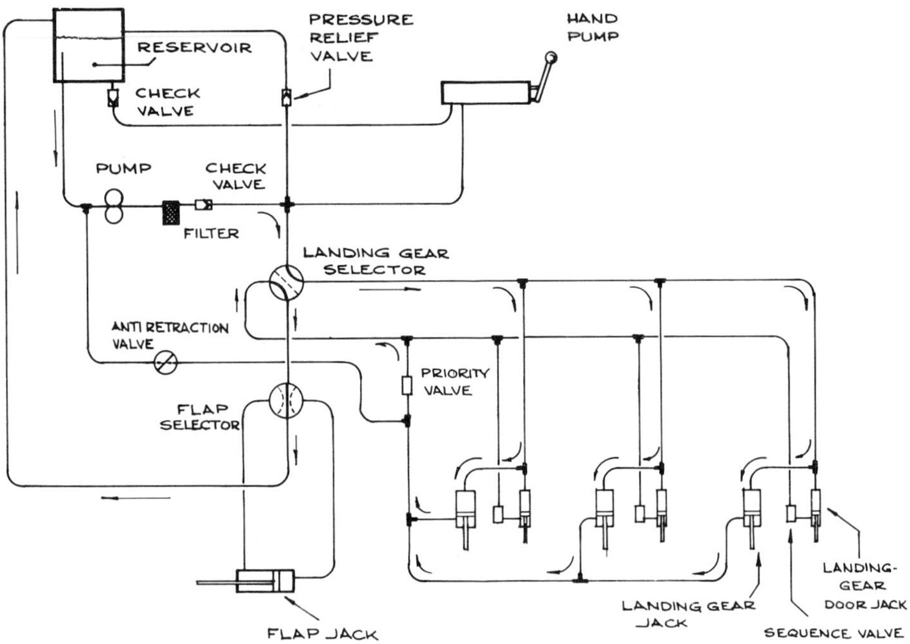

Fig. 9.19 Landing gear down.

pressure. As soon as the landing gear has completed its cycle and the selector has returned to neutral the flaps will move and complete their cycle.

There are several components in the system that are only used in an emergency or to provide a safety interlock. The anti-retraction valve is mechanically linked to one of the main landing gear oleo struts and is open when the gear is compressed at all from its fully extended position as, for example, when the aircraft is on the ground. Should the landing gear be selected to the up position the pump pressure will be returned to the reservoir without operating the actuators. The operation and adjustment of this valve should be checked each time any maintenance is carried out on the landing gear.

In the event of a failure in the engine-driven hydraulic pump the emergency hand pump may be used to provide the hydraulic pressure necessary to operate the systems. This draws fluid from the bottom of the reservoir through a strainer and check valve. The pressure from the pump is then fed into the system for use as required. The main system check valve prevents the pressure from the hand pump being dissipated through the failed engine-driven pump.

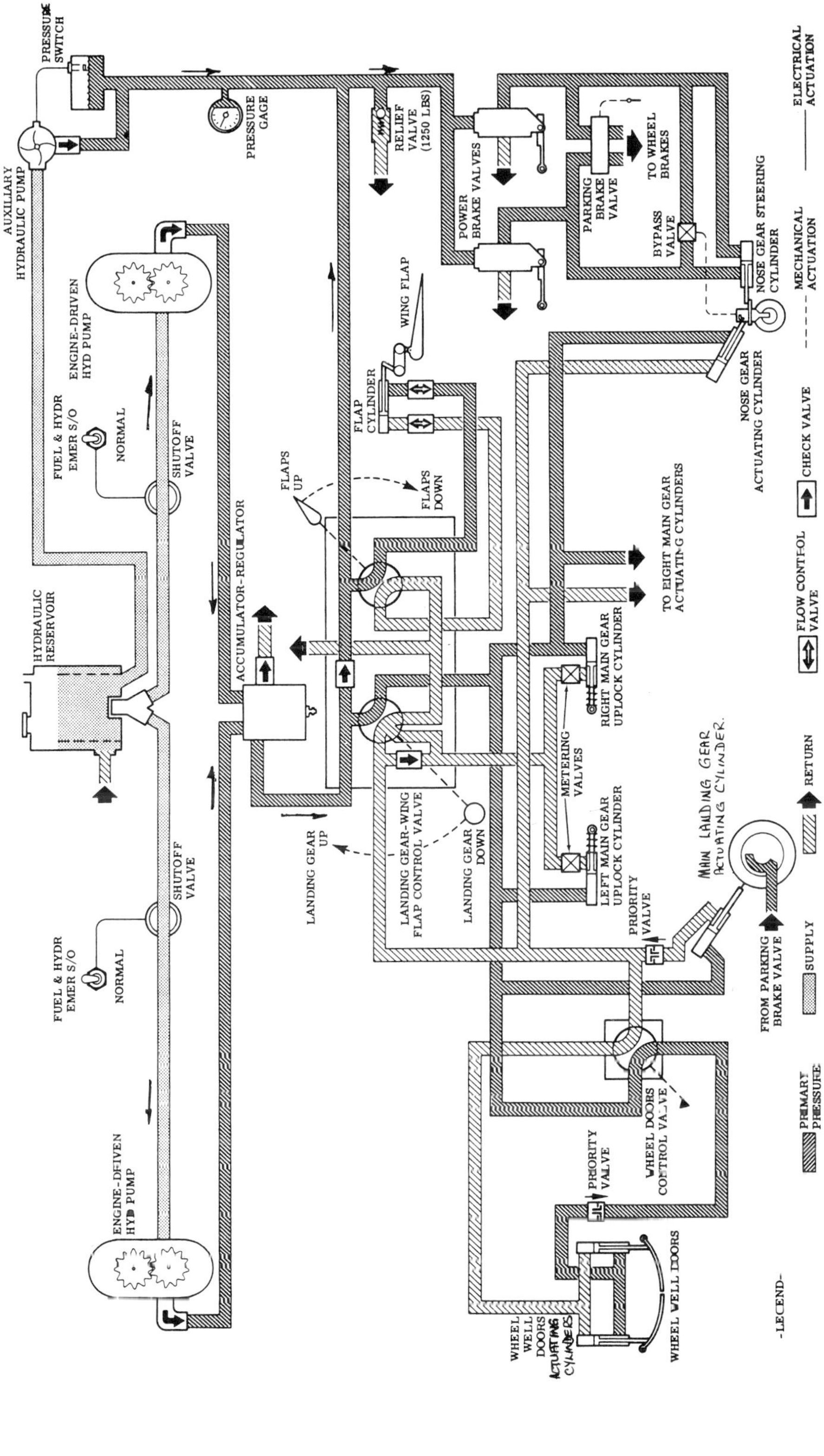

Fig. 9.20 Closed centre hydraulic system.

Closed centre hydraulic system

The operation of individual systems, such as landing gear and flaps, is similar to that of the open centre system described. The components are not usually grouped in a powerpack and the pump is not unloaded through the selector valves but through a separate unloader valve. The system illustrated also incorporates the filter in the reservoir and combines the unloader valve with a hydraulic accumulator and a pressure relief valve. Also the system is shown with two engine-driven pumps (fig.9.20).

The unloader valve is designed to keep the pressure in the main hydraulic system within certain limits and to allow the pump to idle when its output is not required. The piston type accumulator keeps pressure on the system and provides a 'cushion' in the system by absorbing the pulsations caused by the intermittent flow from the

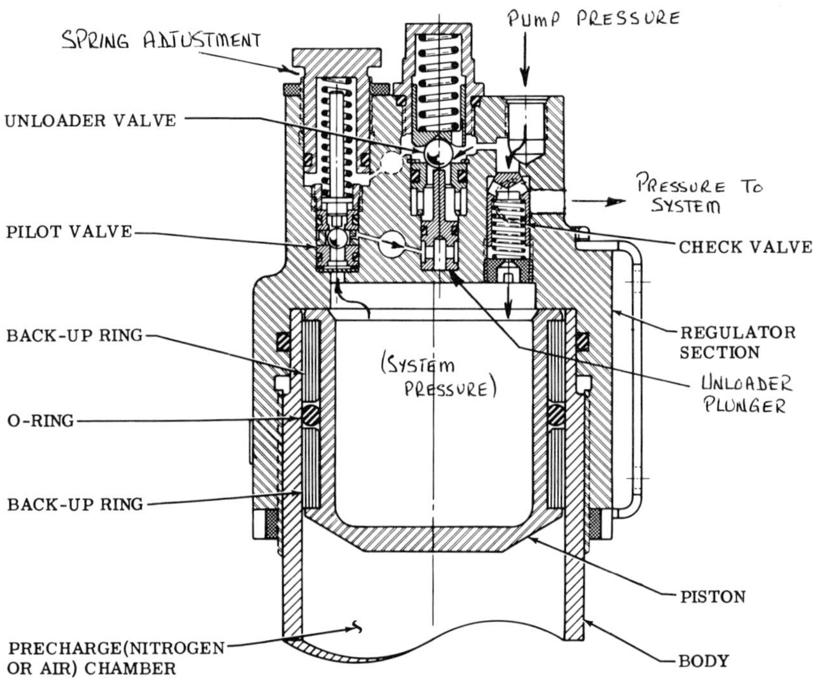

Fig. 9.21 Accumulator/unloader valve.

pump (fig.9.21). When the system pressure falls below the predetermined value the pilot valve is kept on its seat by the adjustable spring and so the unloader valve is kept closed by its spring and the pump supplies the system. When the system pressure reaches its maximum value the pilot valve is lifted and system pressure is able to lift the unloader plunger and open the unloader valve. This allows the pump output to return to the reservoir. A system check valve prevents loss of the system pressure when the pump is unloaded. A pressure relief valve limits the maximum value of the hydraulic pressure in the event of thermal expansion or failure of the unloading valve. The accumulator must be pressurised with nitrogen to the correct pressure. Nitrogen is used rather than air in order to reduce the chance of internal corrosion in the accumulator cylinder.

Helicopter hydraulic servo system

Many light helicopters have completely manual flight control systems but it is increasingly common, particularly among turbine-powered helicopters, for some form of hydraulic servo assistance to be fitted. A servo actuator is basically a hydraulic cylinder with a small control valve attached to it that is operated by the pilot's controls. The mechanical linkage to this valve is such that movement of the cylinder, that is extension or retraction, is fed back to the valve and so stops the movement when the desired travel has taken place. This action is called feed-back or follow-up. Such servo actuators may be fitted to both cyclic axes, the collective and directional controls, although there is a move away from the last named. The directional control forces in most light helicopters are not really high enough to justify a tail rotor servo.

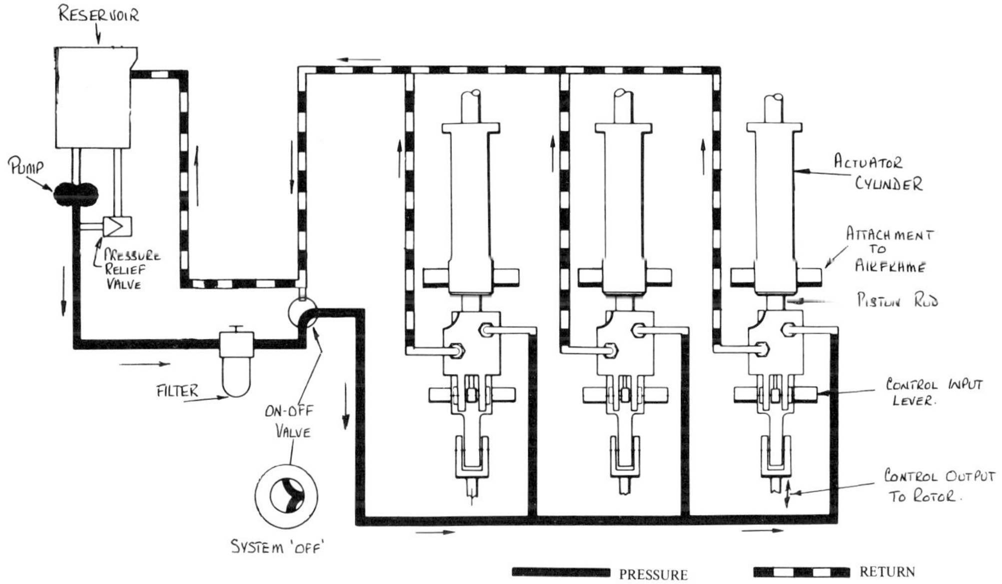

Fig. 9.22 Helicopter hydraulic system.

The external components of the helicopter hydraulic system are very similar to those employed in a hydraulic system designed for an aeroplane (fig.9.22). The pump is usually driven by the main rotor transmission so that powered control is maintained in the event of an engine failure. Having said that there is one helicopter that has had an engine-driven pump for a good many years. From the pump the fluid passes through a filter to a control valve. This valve, which may be electrically or manually operated, bypasses the pump output back to the reservoir if the pilot desires to switch off the hydraulic servo assistance. From the control valve the fluid is directed to the servo actuators which return the fluid to the reservoir after use. More recent helicopters have taken their cue from their aeroplane counterparts and have combined the various supply components into a single hydraulic powerpack. One distinct difference from the systems used in aeroplanes is the operating pressure, which is controlled by the pressure relief valve. Operating pressure for typical systems vary from approximately 250 to 800 p.s.i. (17.25–55.0 bars).

The servo actuator requires a little closer attention if a clear understanding of its functioning is to be gained. When the control stick is moved the linkage moves the spool valve lever a small amount, limited by an elongated hole in the piston rod (fig.

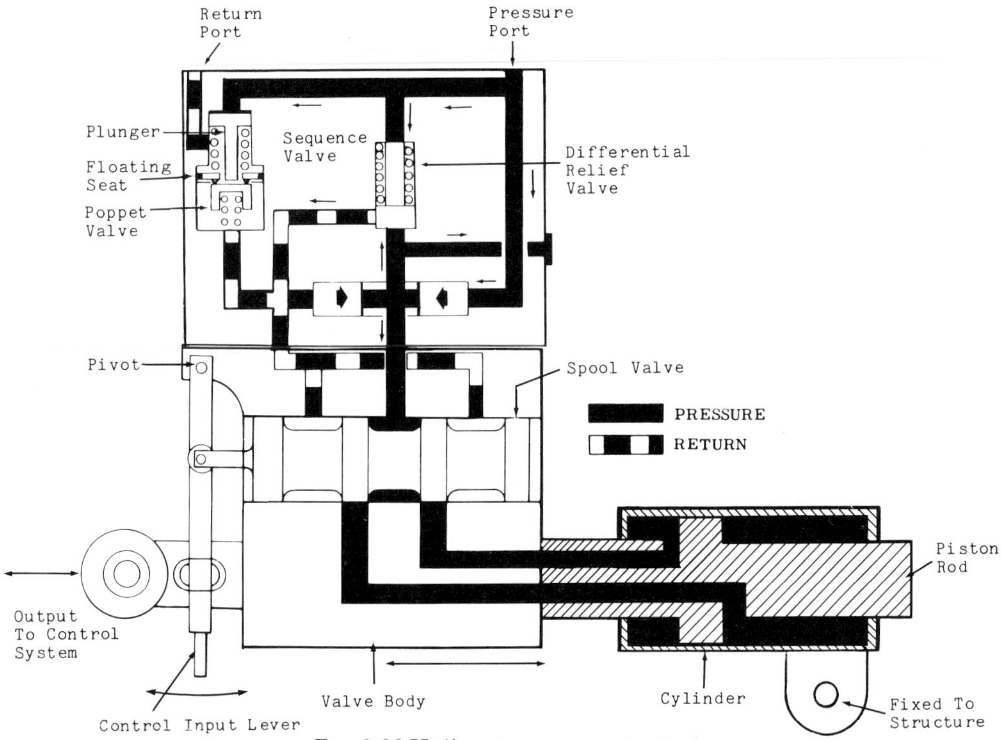

Fig. 9.23 Helicopter servo actuator.

9.23). This movement of the valve opens the ports to admit fluid under pressure to one end of the cylinder and to release trapped fluid from the other end to return to the reservoir. When movement of the control stick is stopped the spool valve lever becomes stationary, although the valve body is still moving under the hydraulic pressure in the cylinder. As the body moves so it closes off the valve ports and the actuator stops moving, having reached its new position as required by the pilot. Once the actuator reaches this position hydraulic fluid is trapped in the cylinder on each side of the piston, thus providing a hydraulic lock that holds any loads imposed by the rotor. This is the normal operation of the servo actuator.

Consideration must also be given to the operation of the servo under emergency conditions, for example in the case of loss of hydraulic pressure. Whilst there is operating pressure at the inlet port the plunger in the sequence valve holds the poppet valve off its floating seat and so permits the return flow of fluid to the reservoir. With the loss of hydraulic pressure the plunger is pushed up by the spring and the poppet valve thus closes the return port. Subsequent movement of the spool valve by the pilot will permit the oil to circulate within the actuator as it is moved by the force applied by the pilot. As soon as the spool valve is centred there will once again be a hydraulic lock in the cylinder that will carry the control loads from the rotor. Thus the fluid is retained in the actuator even if there should be a complete failure of one of the flexible hoses in the hydraulic system. The sequence valve also serves to relieve thermal pressure build-up should this occur whilst the system is inactive. If internal pressure does build up the floating valve seat is pushed up compressing the upper spring. The poppet valve is prevented from following it by an internal stop in the valve shown as a line on the drawing. The differential relief valve has system pressure on both sides of it during normal operation and only on the lower side during emergency operation and it serves to relieve any pressure build-up that could occur when the spool valve is centred with both return ports closed.

Electro-hydraulic powerpacks

In recent years a number of light aircraft have been produced with hydraulic systems that do not depend on an engine-driven pump for the source of power. The advantage in a single-engined aircraft should be obvious in that power is still available in the case of an engine failure, a time when a pilot has his hands full without adding the complications of emergency landing gear extension as well. Generally speaking this type of unit has only been applied to the landing gear, systems such as flaps being electrically operated. The powerpack is similar to other hydraulic powerpacks in that it incorporates the reservoir and control valves for the system and additionally an electric motor driving a hydraulic pump. There are two distinct designs of powerpack which may be readily differentiated by the means employed for 'up' and 'down' operation. In one design the motor and hydraulic pump are reversible, the direction of rotation being controlled by the pilot's selector switch. In the other main design the pump and motor always rotate in the same direction and the hydraulic fluid is routed in the correct direction by means of solenoid operated valves sequenced by the pilot's selector switch and microswitches on the landing gear. In both types the motor is stopped by a pressure switch when the maximum hydraulic pressure is reached. An additional servicing item with this type of powerpack is periodic inspection of the brushes in the electric motor, a job which may be accomplished with varying degrees of ease depending on the type of unit and where it is installed in the aircraft.

Fault diagnosis in hydraulic systems

There is no possible way in which this section can cover all the possible faults that may be experienced in specific hydraulic systems but there are some pointers that can be given that may lead in the right direction. The basic principle in all fault diagnosis is to isolate the problem and then to eliminate the good items from that area of the system. It goes without saying that it is impossible to isolate the problem in any but the simplest system without a thorough knowledge of the operation of the system. Collect the known data, the symptoms from the flight crew and your own observations, and then sit down with the schematic diagram of the system and work logically to isolate the fault. Do not make a 'stab in the dark' at a solution!

Hydraulic systems can be divided into two basic areas – power and actuation. The power system will include both the main system and the emergency system. If a fault occurs when operating on the main system which does not occur when operating on the emergency system the fault must be located in that part of the main system that is not common with the emergency system. For example the fluid level for the emergency system is taken from the reservoir at a lower level than that for the main system, so a low fluid level could be the fault. Check it and, if possible, eliminate it from the list. Start with the easiest things first and eliminate them before moving on to the more complex and time-consuming items. Other items that could be at fault are the pump and its connecting pipework. If there is a separate pressure relief valve for the pump system it could be stuck open but if it is common to both systems it cannot be the problem. If only one actuation system fails to operate, for example the flaps, and the others are operating normally then the trouble must lie in that system area. There are a number of symptoms of which we should be aware since they are common to most systems but they are not treated in any special order of merit and are passed on as a practical aid to fault diagnosis.

If there is no pressure indicated on the system pressure gauge yet there is a return flow into the reservoir it is unlikely that the trouble lies with the pump. It is more likely that the system is short circuited by a pressure relief or unloading valve being stuck in the open position.

Where the system pressure is higher than normal it could well be caused by the unloading valve failing to unload the pump and the system pressure being limited by a relief valve in the system. Make sure that you know the value of each valve setting and use this information in arriving at your diagnosis. Incidentally if a relief valve is taking the full pump flow and pressure it will often make a buzzing sound and get quite warm.

A loud hammering noise in a system fitted with an accumulator is almost certainly due to a lack of, or insufficient, air charge in the accumulator.

Pump chattering and subsequent overheating can often be caused by air in the line. Air could enter the system by the reservoir level being low or through an air leak in the suction line between the reservoir and the pump.

Slow actuation of one unit could be caused by internal leakage in the actuator or in the control valve. This leakage flow will tend to raise the temperature of the offending unit given a little time. Internal leakage can be confirmed in an actuator by moving the piston under pressure to one end of the cylinder. With the pressure still applied disconnect the return line and observe the open port for any sign of hydraulic fluid leakage. If all systems are slow then suspect that the reservoir level is low.

Spongy actuation of a system is usually due to air in a system and, predominantly, this is a problem associated with single-acting systems such as wheel brakes. A double-acting system is self bleeding in most circumstances, it only being necessary to cycle the actuator several times to clear any air trapped during the change of a component. Air in the line to a pressure gauge will cause a totally different symptom with the pointer fluctuating so rapidly as to form a blur. This may be dealt with in much the same manner as bleeding the air from a radiator in a central heating system. Wrap a clean rag round the fitting on the back of the gauge and slacken the nut slightly with the pressure applied to expel any air and then retighten.

Another problem that may be experienced is incorrect sequencing of landing gear and hydraulically operated doors. A study of the open centre system illustrated earlier in this chapter will show that there could be a fault in the priority valve if all three gears are affected, or in a sequence valve if only one gear is affected. In other systems the gear and doors are sequenced electrically and the rigging of the microswitches on the landing gear is particularly critical. These microswitches may be protected against the ingress of moisture by coating them with a suitable silicone compound.

If nothing moves in a system with an electro-hydraulic powerpack start by checking that the electric pump operates properly and there is sufficient voltage available at the terminals of the pump. Low voltage, for example caused by a high resistance connection, or worn-out brushes in the motor, will reduce motor speed to a point where the pump will be unable to properly supply the system. When rigging the landing gear fitted to this type of aircraft always ensure that a ground power unit is used and not the aircraft battery, for the typical powerpack will draw 30–50 amps.

If a landing gear fails to retract and all else appears normal the fault may lie in the anti-retraction (safety) switch or valve. Check it before dismantling the rest of the system.

A helicopter servo-actuator that is jerky or overcorrects is often caused by a binding spool valve in the actuator. 'Motoring' is a condition in which the actuator tends to move persistently in one direction even after control input movement has stopped. This too can be caused by a sticky spool valve but also by internal leakage across the valve.

'Chattering' in a helicopter system is often caused by misalignment of the cylinder, the hoses or the control linkage. The cylinder must be completely free to take up a natural position on the spherical end bearings under all conditions.

Excessive feedback of control forces to the pilot's stick can be caused by low hydraulic pressure or by an improperly rigged rotor. The latter will usually produce other symptoms that will enable the fault to be isolated to either the hydraulic system or the rotor system followed by appropriate corrective action.

Adjusting the hydraulic system

With the aircraft on jacks it is possible to carry out a simple functioning check of the systems using the emergency hand pump. However it should be borne in mind that some manufacturers do not permit landing gear retraction using the hand pump unless the weight of the gear is supported and some forbid it completely. For any landing gear rigging and any hydraulic system adjustments a hydraulic rig is essential. Unless the full flow is available as well as the correct pressure it is not possible to be sure that relief valve settings are correct. Furthermore some aircraft have a reputation for being correctly rigged on the hand pump and yet misbehave as soon as they take to the air. This is particularly true of aircraft with steel spring main landing gears that are fitted to the fuselage. The speed of retraction is critical in making the necessary up and downlock adjustments and if these are incorrect under real conditions then the doors will malfunction as well.

A hydraulic rig may be purchased or may be manufactured locally and includes an electrically driven hydraulic pump, a reservoir, a filter, a bypass valve and a pressure gauge. It may contain a lot more but this is about the workable minimum. In use the rig is connected to the test connections on the aircraft or, where these are not provided, in place of the engine-driven hydraulic pump. The bypass valve on the rig is opened wide and the electric pump switched on. The bypass valve is closed slowly until the required pressure is indicated on the gauge and then the maintenance work on the aircraft can proceed. One advantage of using the rig is that by use of the bypass valve the speed of operation can be slowed down in order to study any particular phase of an operating cycle. Some rigs incorporate a hand pump or screw jack to permit proof testing at 1½ times working pressure but this function can equally well be achieved by a separate hand pump. One point about the above type of rig is that it will be necessary to connect the vent of the reservoir on the aircraft to the resevoir of the rig in order to contain what would otherwise be spillage due to the volume of fluid in the rig. On completion of the operations on the aircraft the aircraft reservoir should be full due to the action of the rig. If the rig is left connected to the aircraft for a preriod of time without operation there can be a tendency for the fluid in the aircraft, which is at a higher level, to syphon back to the rig and this loss may only be discovered during the first flight after maintenance. Always disconnect the rig as soon as the work is finished.

CHAPTER 10

Helicopter Rotor and Transmission Systems

As we have progresssed through the various chapters of this book we have tried to emphasise that the technology of the aeroplane and the helicopter have much in common even though the physical application may vary on occasions. In the helicopter the rotor system is probably the most obvious difference between the types of machine and yet the principles by which it develops lift are identical. In fact helicopters, along with other types of rotorcraft, are sometimes known as rotary wing aircraft in recognition of this fact. Unlike the fixed-wing aircraft the aerofoil section is often symmetrical because of the different operating conditions.

Rotor blades

There have been many different methods of blade construction over the years, starting with a steel tubular spar upon which was built a wood and fabric

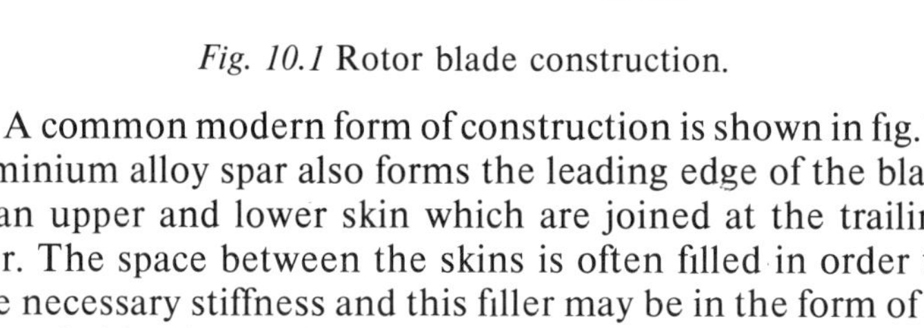

Fig. 10.1 Rotor blade construction.

construction. A common modern form of construction is shown in fig.10.1 where the extruded aluminium alloy spar also forms the leading edge of the blade. Bonded to the spar are an upper and lower skin which are joined at the trailing edge by an extruded filler. The space between the skins is often filled in order to provide the blade with the necessary stiffness and this filler may be in the form of an aluminium honeycomb bonded in place or in some blades the space is filled with an expanded polyurethane foam. The latest designs of rotor blade embody a composite construction in which continuous rovings are wound round the blade root attachment bushes, along the length of the blade and back again to form the main structural member. The blade is then built up with polyester resin, fillers and reinforcement matting to the required shape and cured under heat and pressure. The rovings and reinforcement matting may be glass fibre but more work is being carried out utilising carbon fibre.

Whatever the method of construction new blades are carefully trimmed and balanced and tested dynamically against a master blade so that they all fly the same, within a narrow tolerance. The aerodynamics of the blade and its profile are critical to its proper operation with a minimum of vibration, and at the same time its structural integrity must be beyond doubt in view of the fact that the axial load in a typical main rotor blade for a light helicopter may be 2½ tons or more. All these considerations must be taken into account in a structure that must be as light as possible if the helicopter is to lift any commercial load. In view of the foregoing it can be seen that the maintenance of the rotor blade must be of the highest order if safety is not to be compromised. Even the smallest defects must be investigated and corrected.

Maintenance, in general terms, consists in keeping the blades clean and very carefully inspected for any sign of damage, erosion or corrosion. Since the blade behaves as a cantilever beam both the shear force and the bending moment will be highest at the blade root and therefore additional care should be taken when inspecting the root area. In the root, or grip, area will be found overlapping and tapered doublers that are designed to spread the load and these are usually bonded in place. The edges of the doublers should be inspected for any evidence of voids in the bonding using a 0.0015 in. (0.04 mm) feeler gauge to probe any suspect areas. The sealing around the edge of the doublers is often made with a RTV (room temperature vulcanising) silicone type of compound. It is important that this seal is maintained in order to exclude moisture which could cause corrosion or frost damage. The leading edge is subject to erosion which gives the leading edge doubler a sandblasted appearance. The severity of any defect that is allowed to remain must be ascertained from the manufacturer's maintenance manual. Generalisations will not do! Where repair is permitted the defect must be blended out to a shallow, smooth depression using needle and riffler files. All file marks must be removed and the repair polished using 600 grade 'wet and dry paper'. The area should then be checked for cracks using the dye penetrant method. Finally it is necessary to restore the surface protection. Once the blade has been cleaned and inspected and any surface finish restored it is usual to protect the blade by wax polishing. Unless specified otherwise in the maintenance manual any good quality automotive wax polish should be satisfactory provided it does not contain silicone. A cautionary note: when repainting damaged areas of the blade keep the amount of paint applied to a minimum or the balance of the rotor will be upset. This is particularly true of the tail rotor blades which rotate at a higher speed. Whilst this section has been written with main rotor blades in mind the same general considerations apply to tail rotor blades as well.

Main rotor heads

Basically main rotor heads for this class of helicopter can be classified as rigid, semi-rigid or fully articulated. In the rigid head the blades are attached rigidly to the central hub and cannot flap or drag. The only freedom allowed is that of pitch change. In the semi-rigid rotor the blades, usually two only, are rigidly attached to the hub which in turn is pivoted to the top of the mast. This arrangement allows the rotor to see-saw as a unit as well as allowing the individual blades to change pitch. In the fully articulated rotor the hub is rigidly attached to the mast and each blade is free to change pitch, flap and drag. To explain the point a little, flapping is the vertical movement and dragging is the horizontal movement of a blade about hinges that are part of the rotor hub. Some more recent helicopters use rotor heads in which the hinges are replaced by elastomeric bearings which give a limited degree of movement in any direction and yet others use a one-piece forging that can flex in the flapping direction. Flexible glass reinforced plastic rotor heads have also been used successfully. The fully rigid head is very simple and does not require explanation but we will give some time to the other two head types.

The semi-rigid rotor

In its most common form this is a two-bladed rotor in which the blades are rigidly attached to the hub (yoke) and the whole assembly is free to tilt about a trunnion attached to the top of the mast. This type of rotor has the advantage of simplicity as well as avoiding complications from Coriolis effect. Coriolis effect is not present in this type of rotor since the centres of gravity of the blades always remain equidistant from the centre of rotation, that is the pivot point on the trunnion (fig.10.2). The

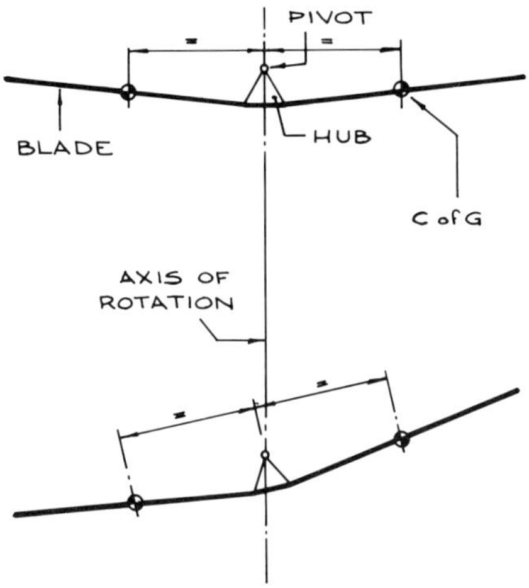

Fig. 10.2 Semi-rigid rotor theory.

trunnion is splined to the top of the mast and the pivot is provided by two plummer (pillow) blocks bolted to the upper surface of the hub (fig.10.3). Two angular contact ball bearings provide the necessary bearing arrangement in each plummer block. Each blade is mounted in a grip (sleeve) by a single hollow bolt which also acts as a container for lead shot for balancing purposes. The blade is prevented from pivoting about this bolt by a drag brace or a blade latch, both of which are adjustable to provide the correct blade alignment. The grip is mounted on the hub spigot by means of two roller bearings, both being lubricated by an integral oil reservoir (fig.10.4). Grease lubrication is provided on some heads as an alternative to oil. The centrifugal loads imposed during rotation are absorbed by a tension-torsion strap or, in some heads, by a ball thrust race. The tension-torsion strap consists of two steel bushings that are spaced a fixed distance apart and are then wound with a continuous solid strand of steel wire. The whole assembly is then encapsulated in a plastic resin. The finished component will permit twisting motion but will resist any longitudinal movement, that is, in tension. One end of the tension-torsion strap is anchored to the hub and the other to the grip. The control input rods are connected to pitch change horns on the grips, thus controlling the rotation of the grips and the pitch change of the blades.

Whilst a number of helicopters use the simple head described, others have variations that express the preferences of different manufacturers or the age of the

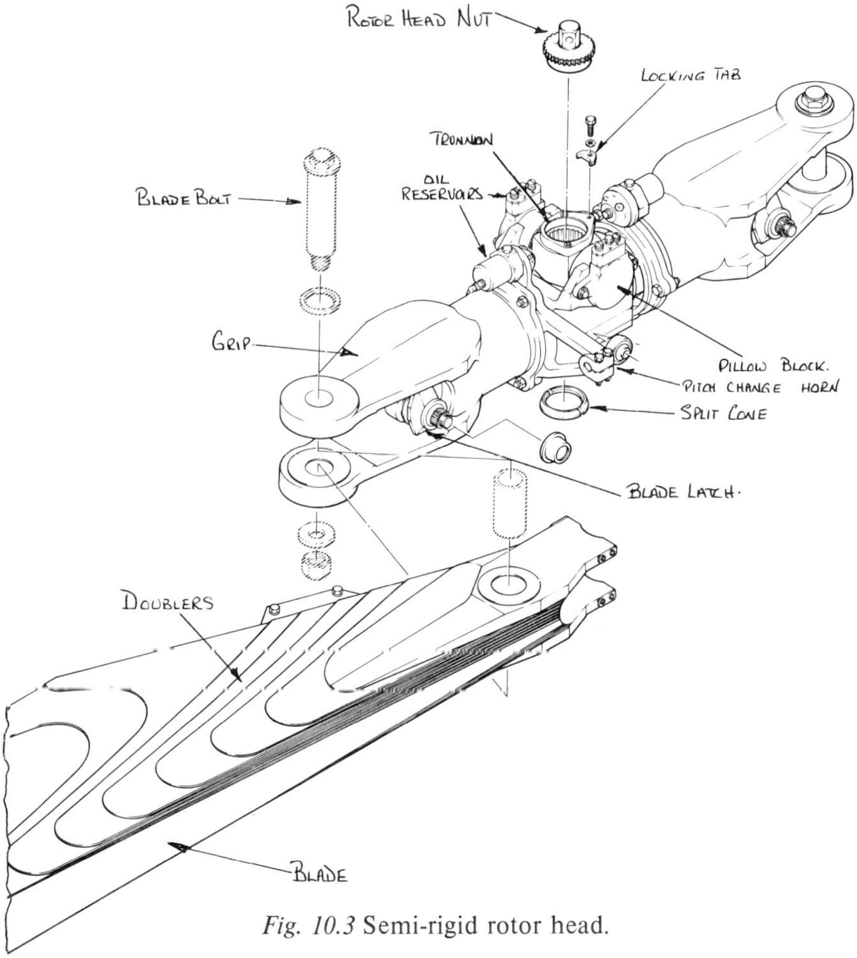

Fig. 10.3 Semi-rigid rotor head.

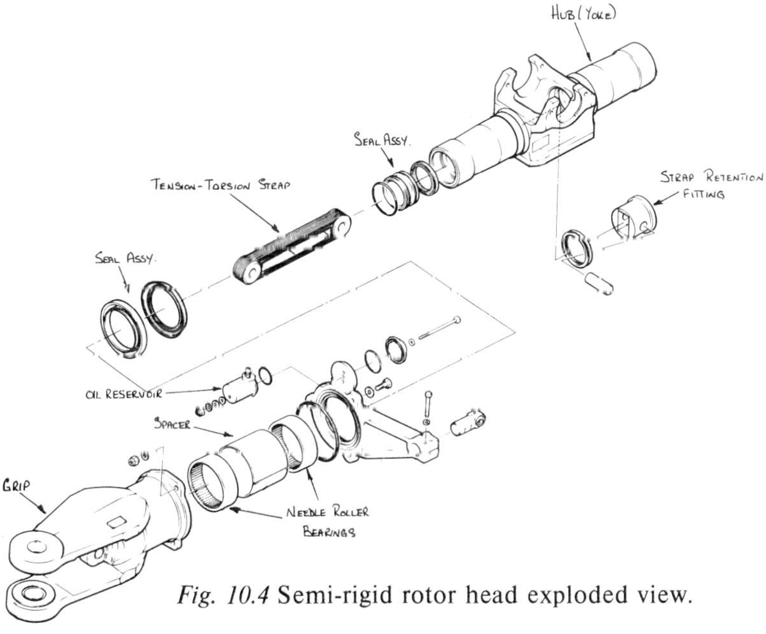

Fig. 10.4 Semi-rigid rotor head exploded view.

design. Some heads are mounted on a full gimbal and thus free to pivot in two directions. In some designs the control inputs change the pitch of two aerodynamic paddles mounted on the hub and the resulting change in lift developed by the paddles tilts the hubs, with the blades, cyclically. The paddles are mounted at 90° to the span of the blades in order to compensate for phase lag.

Another manufacturer mounts a pivoted stabiliser bar at 90° to the blade span and this bar takes up a plane of rotation in the same manner as a gyroscope. If the helicopter or the rotor are disturbed from their steady attitudes the stabiliser bar will remain in its original plane of rotation. Pivoted on the stabiliser bar are mixing levers which receive the control inputs to the rotor and connect to the individual blades by means of short rods. Any change in the relationship between the mast, or the rotor, and the stabiliser bar will result in a control input to the rotor to correct the disturbance. Such a system provides an effective stabiliser but also resists inputs by the pilot, resulting in a helicopter that is so stable as to be unmanoeuvrable. The behaviour of the stabiliser bar can be modified by connecting it to the mast via two rotary hydraulic dampers. These permit the control input to be effective immediately but allow the stabiliser bar to take up its plane of rotation again over a period of some 4-6 seconds. Should a damper become stiff in operation (known as a 'hard' damper) there will be a harsh and sudden control response. Conversely, should it lose its damping ability altogether (a 'soft' damper) the stabiliser bar will exert its authority and the pilot will have little cyclic control.

Inspection and maintenance of a semi-rigid head must include its lubrication. Oil and grease seals occasionally leak and can usually be replaced, although rarely without removal and dismantling of the rotor head. The hub and grips should always be carefully inspected for cracks, using the maintenance manual as a guide to the most critical areas. Nuts and bolts should be kept at their correct torque. Bearing in mind the catastrophic consequences of failure a serviceable torque spanner must always be used on every fastener on the rotor head.

The fully articulated rotor

This type of rotor has more parts than the semi-rigid rotor since each blade is provided with its own set of hinges to give it freedom to flap or drag, but it does permit the design of a rotor with a larger number of blades. In an effort to simplify this type of rotor more recent designs have used one-piece flexible forgings or slabs of glass-reinforced polyester in place of the flapping hinges. Elastomeric bearings have also been used giving limited degrees of flap, drag and pitch change.

The blade is mounted in a grip (sleeve) as before and the grip in turn is mounted on a spindle, being free to rotate on a stack of ball thrust bearings (fig.10.5). The spindle is forked at its inner end and is pivoted to provide the flapping hinge. This pivot point is itself free to rotate in the horizontal plane to allow the dragging movement. The amount and rate of the drag movement is limited by a damper which may be of a friction or hydraulic type. The flapping movement is also limited as the rotor slows down by centrifugally operated stops. As the centrifugal force decreases with rotor deceleration so springs pull the stops into place under the spindles and restrict their downward movement. Control inputs from the swashplate are routed directly to a control horn mounted on each grip.

Inspection and maintenance of an articulated head is much the same as for any rotor head, except that there is a little more of it. Lubrication of the various bearings

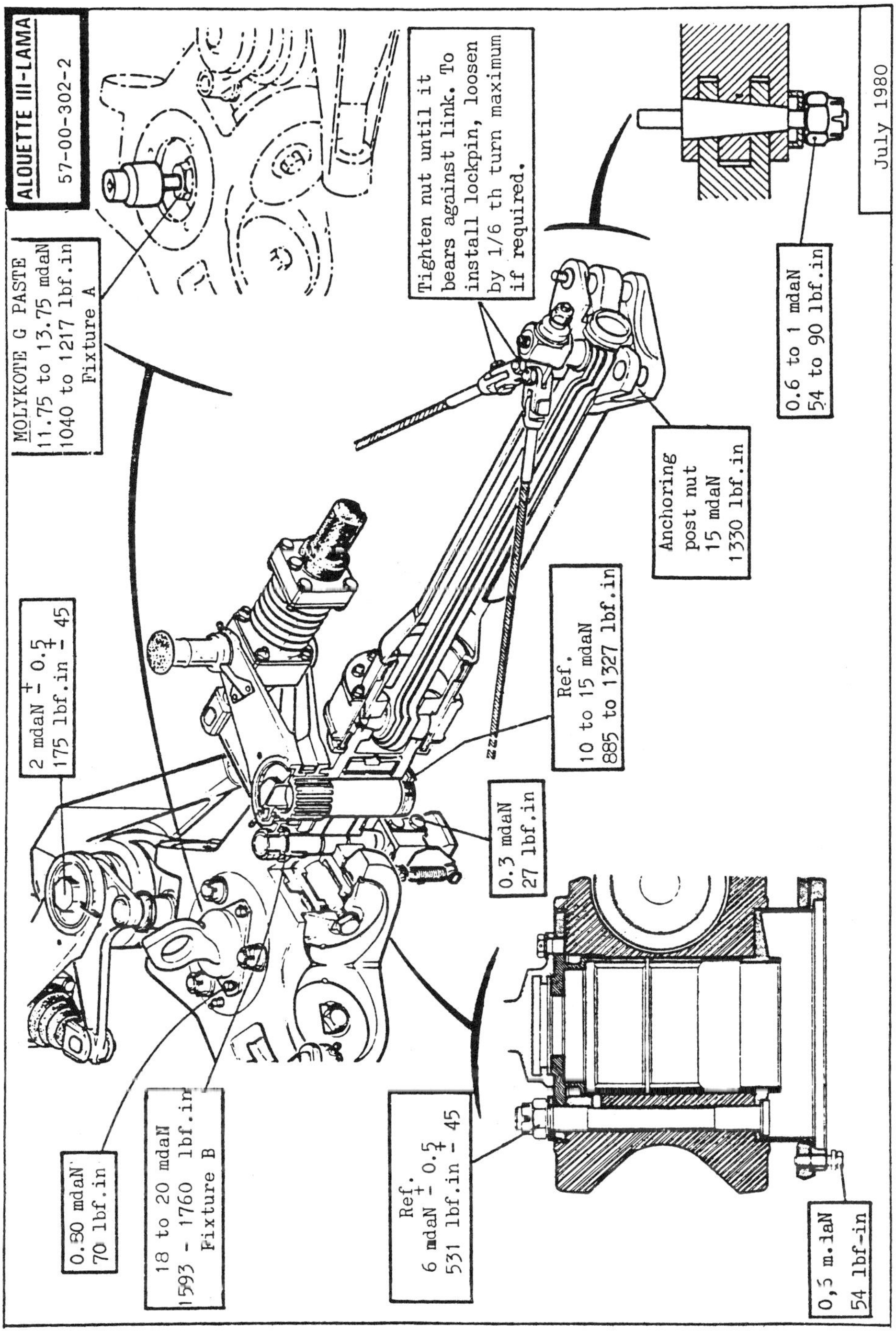

Fig. 10.5 Fully articulated rotor head.

may be accomplished through individual or central oil reservoirs. The drag dampers may be checked for correct functioning by first applying the rotor brake and then moving the blade tip through its full range of movement. The resistance exerted by each damper should be the same and should be constant throughout its travel. A close comparison can be made by using a spring balance to apply force to move each blade and then comparing the readings obtained. Minimum and maximum values are likely to be specified in the maintenance manual. Elastomeric bearings require their own special inspections at each maintenance opportunity. Wear and deterioration of the elastomer will show as small granules of rubber on the surface and is sometimes referred to as 'eraser wear', drawing a comparison with the results of using a pencil eraser. In addition inspection should be made for delamination, checking any suspect areas with a 0.0015 in. (0.04mm) feeler gauge in order to assess the extent of the separation.

Tail rotors

The tail rotor exists to compensate for the torque reaction of the main rotor and to provide the capability of yawing the helicopter. As a result the rotor is much smaller and simpler than the main rotor although it rotates at higher speed. There are probably as many variations as there are manufacturers but only three main types

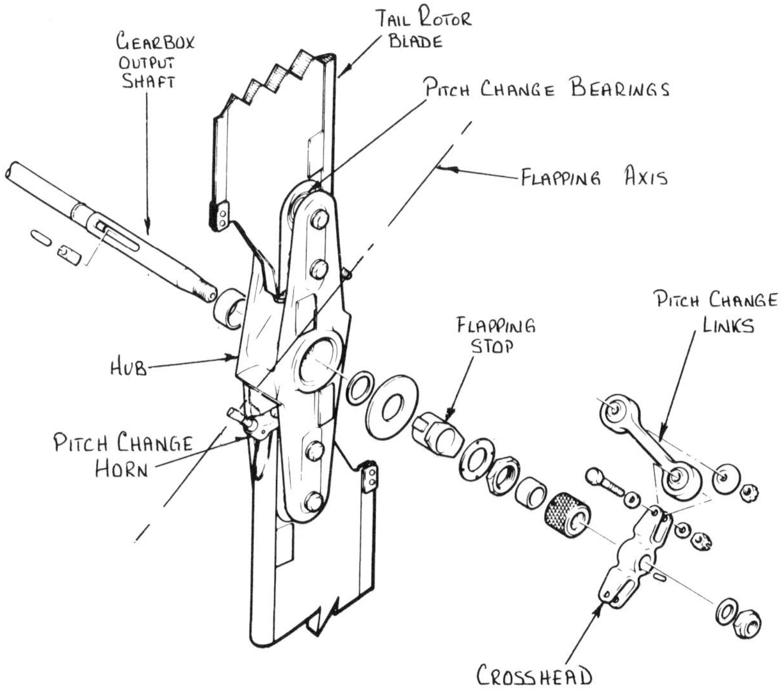

Fig. 10.6 Semi-rigid tail rotor.

will be described. The semi-rigid tail rotor (fig.10.6) is able to seesaw about a trunnion mounted on the output shaft of the tail rotor gearbox. The pivot axis of the hub is not at right angles to the span of the blades and so movement about the axis will cause a change of pitch of the blades. This is known as a delta three hinge and the pitch change may be either positive or negative with outward flapping of the blade. The blades themselves are mounted in grips and are able to change pitch on suitable bearings in the hub. A variant of this type of tail rotor uses a solid slab of aluminium alloy as the hub and all the movements take place on teflon-lined spherical bearings. Individual pitch change rods connect the blades to a crosshead that is mounted on a shaft passing right through the gearbox shaft. The longitudinal movement of this shaft is controlled by a bellcrank or actuator that is connected to the pilot's pedals.

The articulated main rotor also has its counterpart in the tail rotor but the drag hinges are absent. Flapping and pitch change are allowed in the same way as with the main rotor but the lubrication is always grease. Change of the blade pitch is obtained in the manner already described. When the crosshead has more than two arms it is commonly referred to as a spider.

A third variation of tail rotor design mounts opposite blades at the ends of a flexible beam, the centre of which is bolted to the tail rotor hub. The flexible beam may be light alloy or glass reinforced polyester and allows the blades a limited freedom in flapping. Such a rotor must always have two or four blades for constructional reasons.

Rotor adjustment procedures

Whilst every design must have its own particular adjustments that are defined in the maintenance manual there are a number of procedures that are common to more than one rotor. For example, two bladed semi-rigid rotors are aligned and statically balanced before they are fitted to the helicopter and all rotors are tracked after they are installed. So let us take a little time to follow through some typical procedures bearing in mind that the manufacturer's instructions must always be followed to the letter when doing the job on a helicopter.

Rotor alignment

With a two-bladed semi-rigid main rotor where the blades are in a fixed relationship with each other it is necessary to bring them into alignment with each other and the hub. One reason for this is aerodynamic so that the centres of pressure of the blades lie on a line that passes through the centre of rotation of the rotor. The other reason is for static balance in a chordwise sense so that the centres of gravity of the two blades also lie on a line that passes through the centre of the rotor. To achieve this objective the hub is mounted on a stand so that it is free to pivot on its trunnion. The grips are locked in the neutral (0° pitch) setting and the blades are loosely attached. It is important that the grips are not rotated outside their normal range of movement or the tension-torsion straps will be damaged. The blades must be supported at a position about 5 ft.(1.5m) in from the tip so they take up approximately the pre-coned angle. To facilitate subsequent movement of the blades in a chordwise direction (sweeping the blades) it is best that they are supported on a mobile support such as a padded roller skate. A nylon fishing line is now stretched from blade tip to blade tip and secured to the points provided at the tips (fig.10.7). In some cases a small screw is provided which may be unscrewed a turn or two to provide the necessary anchorage. In some blades the points are marked with a rivet and where this is the case it will be necessary to tie weights on to the ends of the fishing line to hold it in place. The blades are then swept in the appropriate direction so that the line passes through reference points on the hub, often the blade attachment bolts. When the line passes through both tip reference points and both hub reference points the rotor is correctly aligned. Since the built-in coning angle of the rotor will place the cord some distance

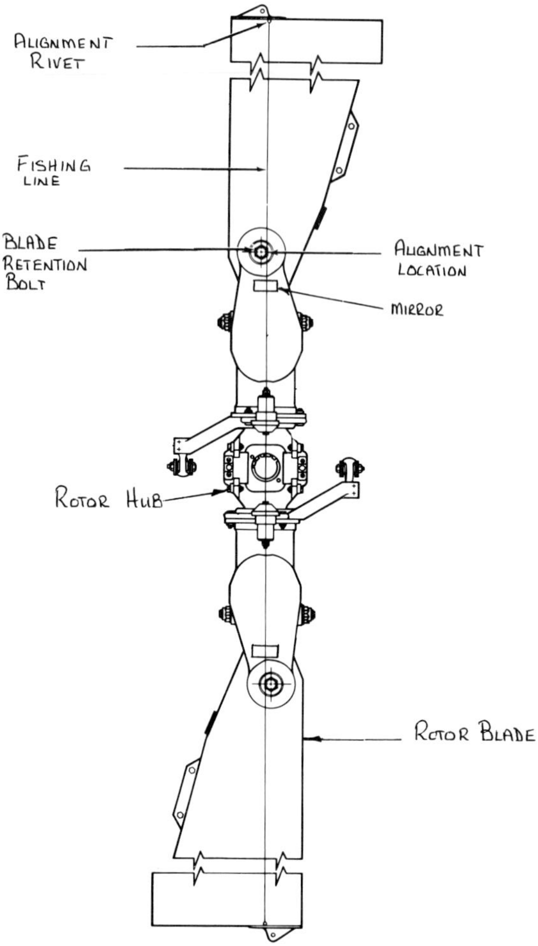

Fig. 10.7 Alignment of semi-rigid rotor.

above the hub some difficulty may be experienced in avoiding parallax error when trying to align the cord with the reference marks. This can be avoided by placing a small mirror just inboard or outboard of the reference points and underneath the cord. As it is viewed from above the cord and its image in the mirror will exactly coincide if the point of view is correct, that is vertically above the cord.

Rotor static balance

So far the consideration has been of a main rotor, but static balance applies equally well to the tail rotor and this will be described separately. To statically balance the two-bladed, semi-rigid main rotor it is mounted as already described in a draught-free area and the blade tip supports are removed, so that it is free to pivot about the trunnion. A bullseye level is mounted on a small fixture on the hub so that it is at the centre of rotation (fig.10.8). Thus any tilt of the hub and blade assembly will be revealed by the position of the bubble within its reference circle. Should the balance be incorrect in a spanwise direction weight is added in the appropriate location, sometimes in the form of lead shot in the hollow blade bolts, sometimes in the form

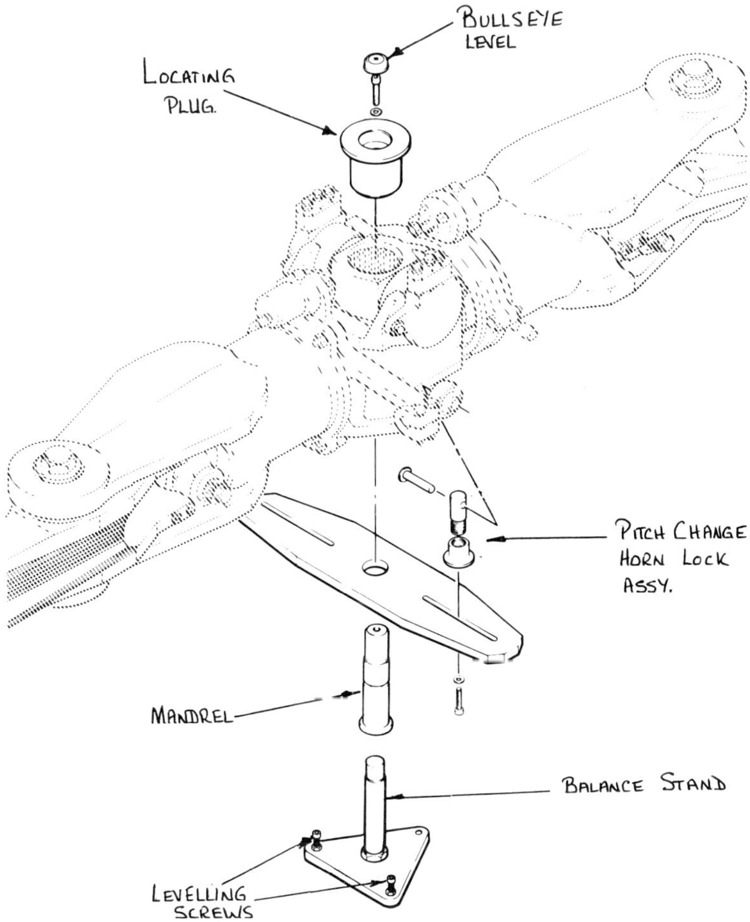

Fig. 10.8 Static balance of semi-rigid rotor.

of washers at some point. Incorrect chordwise balance is corrected by sweeping one blade aft slightly. This is a sensitive adjustment and should not exceed a maximum set by the manufacturer, usually about one flat on the adjusting nut. Never sweep a blade forward after the rotor has been aligned.

With the rotor correctly aligned and balanced all the nuts should be tightened. Using a sensitive torque spanner determine the drag of each nut and then tighten to the recommended torque plus this figure. Make a final check to ensure that nothing has been forgotten.

Balancing a tail rotor is a similar operation but is easier to handle due to its lighter weight. It may be carried out in the manner described for the main rotor, using a suspension balance, and this is often used since there is a minimum of workshop equipment needed. However it may be equally well balanced by mounting the hub on a mandrel and supporting it on two level knife edges. If a two-bladed rotor is placed with the blades vertical any tendency to rotate will reveal incorrect chordwise balance. If the rotor is set with the blades horizontal any tendency to rotate will reveal spanwise unbalance. Due to the different plane of rotation and the different construction of the tail rotor neither lead shot nor the expedient of sweeping the blades can be used to correct the balance. Balance is corrected by adding weights in the locations provided by the manufacturer, often in the form of standard washers.

Rotor tracking

Once the main rotor has been installed on the mast the cyclic and collective control systems are rigged to give the correct travels and blade pitch change. If only the rotor has been disturbed this rigging will probably be limited to the rods from the swashplate to the rotor. The next step is to track the blades in order to bring all the blades into the plane of rotation. The two most commonly used methods of tracking a rotor are the flag method and the stroboscopic method, but essentially these differ only in the method used in measuring the actual difference in track between one blade and the others. This being the case we will describe the flag method (fig.10.9) and then discuss the differences when using electronic means.

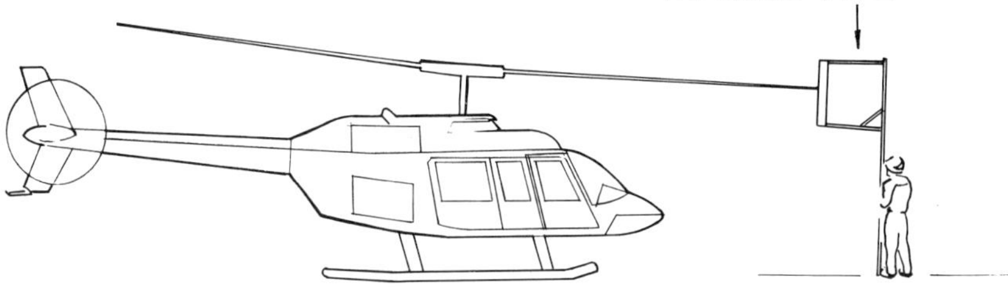

Fig. 10.9 Flag tracking of main rotor.

A frame for a tracking flag may be readily constructed from any light-gauge steel tubing welded together to support a strip of aircraft fabric reinforcing tape. A refinement would be to mount the flag fabric on tension springs to give greater flexibility. Mark each blade tip with a chordwise line using a different coloured wax pencil for each blade. The marks must be in the same position on each blade tip. Face the helicopter into wind on a level, firm surface and mark a position on the ground at approximately the 2 o'clock position in relation to the helicopter and about 1ft. (0.3m) outside the tip path of the rotor blade. The base of the tracking flag should be placed on this mark. The helicopter should be loaded to permit the use of higher power settings without hovering. Any trim tabs should be put to the trail or neutral position unless this is forbidden by the manufacturer.

The helicopter is now operated at 100% Nr (main rotor speed) and collective applied until the helicopter just starts to feel light. It may now be necessary to reduce the rotor speed slightly to some specified figure for checking the track. The person holding the tracking flag should stand with the flag in front of him and his back to the advancing blades. He must be able to see the pilot so that he can receive a signal when to take the track. The base of the tracking flag should be placed on the previously made mark with the frame extending horizontally away from the rotor and the flag should be slowly raised until it approaches the vertical position and the tip path of the blades with the flag tilted slightly in the direction of rotation of the blades. At this point the pilot should check that the flag height is correct, that is that the blades will strike the flag in the centre. When the flag height is correct the flag should be moved slowly into the path of the blades and immediately tipped away when the blades strike the flag. The relative vertical position of the blades will be indicated by transfer of the coloured mark from each blade to the flag. It is good practice to take two tracks prior to making any adjustments in case there was any disturbance due to a gust of wind or slight movement of the controls. The adjustments should be made on the pitch change rods in order to bring the high blade down and never the reverse. This is in order to maintain the rotor speed in autorotation.

Following adjustment as above a further track check should be carried out at 100% Nr and the track adjusted on the tabs where this is permitted. The helicopter should then be test flown throughout the speed range to ensure that the vibration level is satisfactory and to check that the autorotation r.p.m. is correct. If the autorotation r.p.m. is incorrect it will be necessary to adjust all the pitch change rods in the same direction by the same amount. Increasing the pitch of the blades will reduce the r.p.m. and vice versa. With completion of all the adjustments ensure that all the nuts are correctly tightened and safetied. The screwed fittings at the top of the pitch change rods should be sealed by the application of an RTV silicone rubber compound to prevent ingress of moisture. Whenever making adjustments to the rotor always mark the adjusting nuts so that you know how much adjustment has been made and record each adjustment and its effect so that you can be sure you are moving in the right direction.

A slightly more refined method of checking the track that also allows the track to be checked in flight is carried out by attaching a different coloured reflector to each blade tip and then viewing these from the cockpit using a high-powered spotlight powered from the helicopter's electrical system. The difference in track can be estimated visually. A more complicated but more accurate system uses a high-powered strobe flash gun, the flash rate of which is synchronised to the rotor, to view the reflective targets. To simplify the identification of the numbered targets it is possible to reduce the flash rate slightly, when the targets will be seen to spread making them easier to identify, especially on rotors with a larger number of blades. In addition it may be seen whether or not the blades are equally spaced. Unequal spacing in a fully articulated rotor would normally indicate some problem with the drag hinge or damper of the affected blade.

Rotor dynamic balance

In this method an accelerometer is mounted near the main rotor in a lateral position and the output is processed electronically to provide a display on a meter of 'vibration amplitude' (fig.10.10). Strictly speaking this is not amplitude but the peak velocity of the point being measured in inches per second. In addition to the accelerometer a magnetic pickup is used to provide a phase reference, that is one signal per revolution of the rotor. This signal provides a display, similar to a clock face, showing the phase relationship between the rotor and the motion of the vibration and is used to determine the location of the required balance weights. Once the rotor is correctly tracked two readings may be taken from the balancer part of the equipment:

Phase (for example, 12 o'clock, also known as the clock angle)
Amplitude (for example, 0.52 in. per sec.)

The values obtained are now plotted on the nomogram (fig.10.10) from which it is possible to determine the amount of balance weight required and its location on the rotor. The helicopter manufacturer provides certain points on a rotor to which weights may be added or subtracted for balance purposes. It is unlikely that one of these points lies exactly where the weight must be added so a pair of points must be used, adding weights in a manner that will result in the correct effect. The nomogram facilitates the calculation. Each helicoper rotor requires a different nomogram which is calculated specifically for that rotor only. After the rotor has been balanced recheck the track.

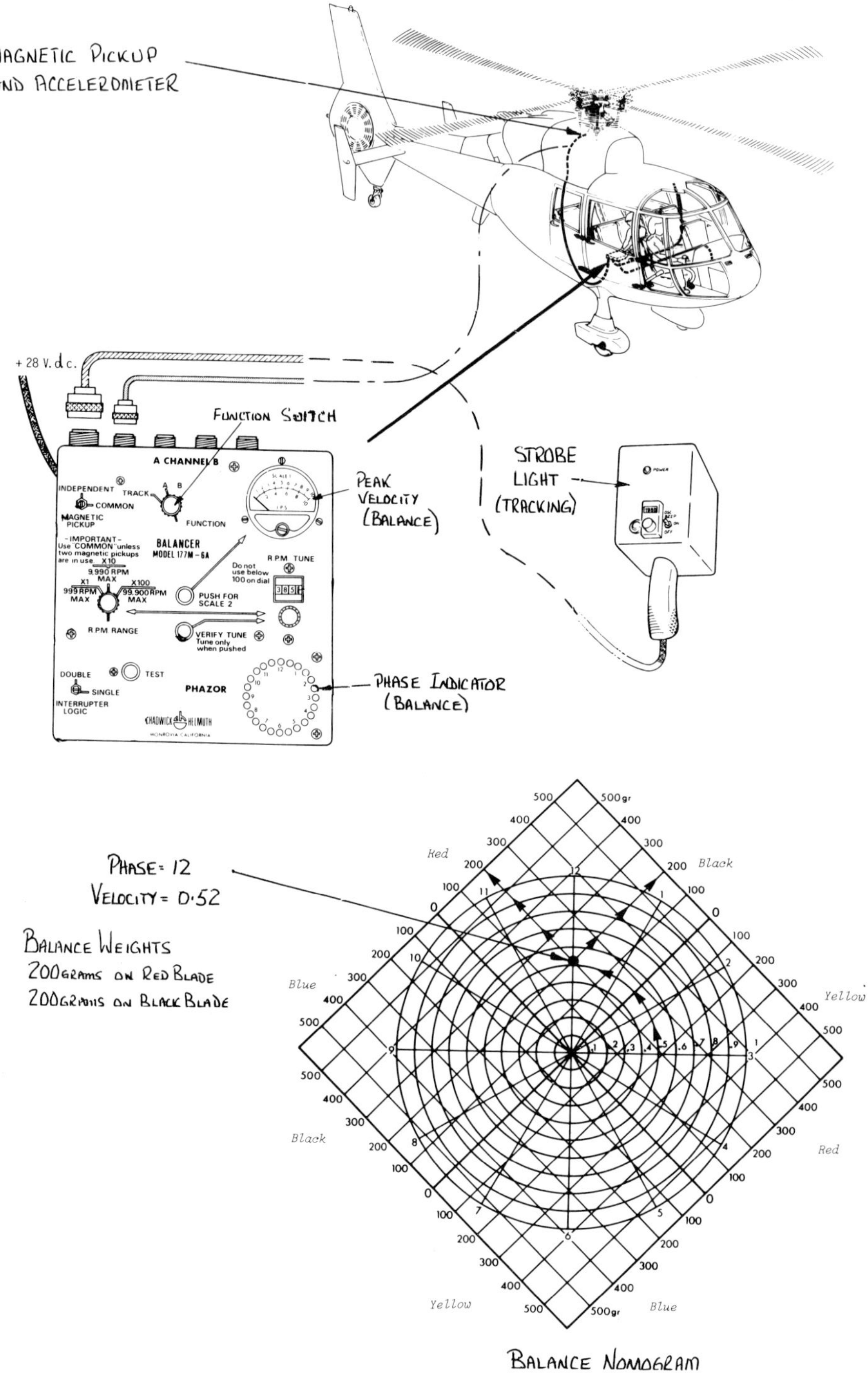

Fig. 10.10 Stroboscopic track and balance equipment

The same procedure is followed for the tail rotor with one or two exceptions. The accelerometer is mounted vertically in order to align with the plane of rotation of the rotor. The magnetic pickup is not used but a reflective target is attached to one of the blades and viewed using the strobe light. The flashing of the strobe light is triggered by the signal from the accelerometer and, as a result, the target will be seen to be stopped at some particular angle. This angle is described by reference to an imaginary clock face and is used in the same way as the phase indication obtained when balancing the main rotor.

Vibration analysis

The same equipment may be used to determine the velocity and frequency of an unidentified vibration. An accelerometer is mounted in the area where the vibration is most obvious and the equipment tuned slowly through the ranges until the meter reading reaches a peak. The frequency may now be read from the tuning dial. The frequency is in a fixed relationship to the rotational speed of the offending component and can therefore be used to accurately locate the source of the vibration. In order to relate the frequency of the measured vibration to the speed of different rotating components an intimate knowledge of the construction of a particular helicopter is required. A vibration of the order of 1/rev. may be related to the speed of rotation as follows:

$$\text{RPM} = \text{f(hz)} \times 60$$

However, not all vibrations are 1/rev. In the case of a condition affecting all the blades in a rotor the frequency would be given by:

$$\text{f(hz)} = \frac{\text{RPM x no. of blades}}{60}$$

In the same way the frequency produced by the meshing of gear teeth may be calculated if the number of teeth on the gear and its speed are known. In this way a table of vibration orders may be produced for a specific model helicopter and many manufacturers publish such a table in the maintenance manual. In identifying the different orders a form of shorthand is often used and some examples should make this clear:

1R = 1/rev. main rotor	5R = 5/rev. main rotor
1T = 1/rev. tail rotor	1E = 1/rev. engine output shaft

The letter M used as a prefix would indicate that the vibration emanates from the meshing of the gear identified. It is usual to quote the vibration orders based on speeds equating to 100% Nr, that is main rotor speed.

In flight a practical test will readily identify a vibration as being either vertical or lateral. If the vibration becomes worse with increasing airspeed the vibration is vertical and the problem is with the tracking of the rotor. A lateral vibration becomes worse with increasing rotor speed and is usually due to unbalance in the rotor system. However to retrack or to rebalance the rotor may not be the solution since the cause may be due to wear in the pitch change rod bearings, a faulty damper or brinelled hinge bearing. Like a good doctor do not merely treat the symptoms but get to, and correct, the root cause of the problem. The test equipment is an aid to fault diagnosis, not a substitute for it.

Helicopter transmission systems

The term transmission system is used to include all the parts of the mechanism that is used to transmit the power from the engine to the rotors and driven accessories (fig.10.11). Therefore this section includes the gearboxes, couplings, shafts, bearings, freewheels and other components that form part of this system. The actual layout varies from model to model and also depends on whether the helicopter is powered by a piston or turbine engine. Typically the drive from the engine passes through a clutch if a piston of fixed-shaft turbine engine is installed, to enable the engine to start without being under load, and then through a freewheel unit that will allow the rotor to continue turning in the event of engine failure and so permit an autorotative descent. The freewheel unit may form part of the main gearbox or part of the drive to it. The main gearbox reduces the speed of rotation to that required by the main rotor and often provides the drive to the tail rotor and various accessories, such as the engine cooling fan, the hydraulic pump and the rotor tachometer.

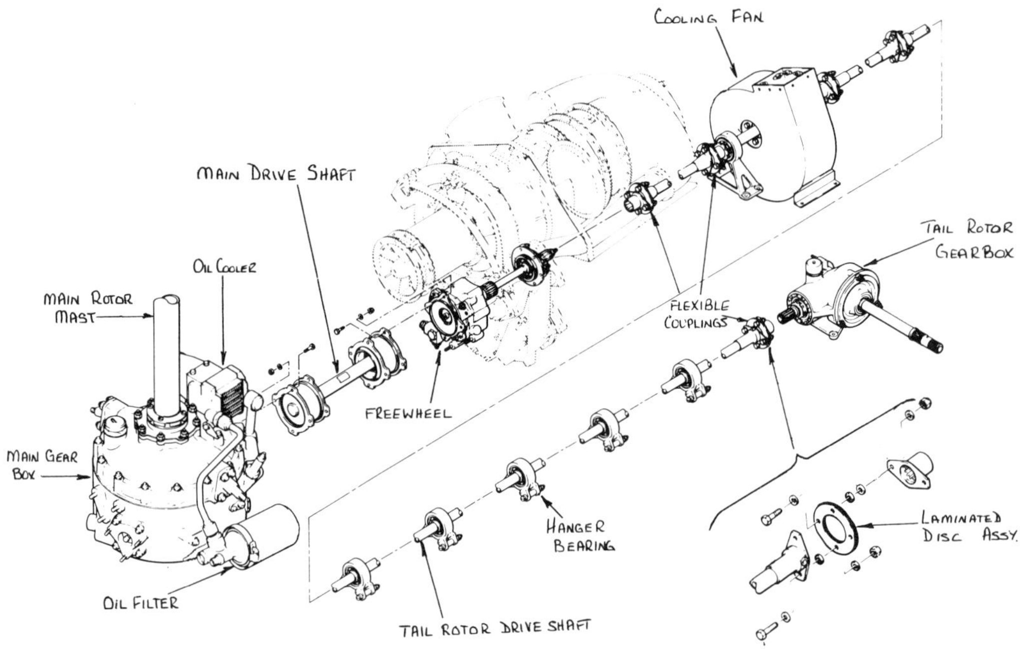

Fig. 10.11 Power transmission layout.

Belt drives

Several light piston engine helicopters use a belt drive between the engine and the gear box input shaft. This has the advantage of cushioning the drive and smoothing out the firing impulses of the four-stroke engine. By means of a movable jockey pulley the belt drive may also function as a clutch, the belts being slackened and allowed to slip for starting. Where multiple belts are used in a drive they must be a matched set and should always be renewed as a matched set. If one belt alone was renewed it would tend to take most of the load and fail relatively quickly. The belts must be accurately aligned with the pulley grooves and correctly tensioned. Alignment between the two pulleys can be checked by laying a steel straightedge along the face of the pulleys or along the grooves. The pulley shafts must be parallel. If the belt tension is too low the belts will slip, generate excessive heat and be weakened. If the tension is too high the tensile strength of the belts may be exceeded. Where a jockey pulley is employed either to maintain the belt tension or to act as a clutch it is important that it is precisely aligned with the driving and driven pulleys. Uneven belt tension and excessive wear, particularly at the edges, can all be caused by incorrect jockey pulley alignment. As with other rubber parts it is important to ensure that the belts are kept free of oil and fuel which would contaminate and deteriorate the belts. Regular inspection should include all the factors already mentioned as well as checking for wear and cracks in the belt.

Clutches

When a clutch is employed it is more usual for a centrifugal type to be installed in which the drive is taken up progressively as the speed is increased. In some the mass is simply the shoes carrying the friction linings but in at least one other example the mass is provided by a container of mercury. The primary check as to the serviceability of a clutch is the time it takes for engagement at a certain speed. This should be checked by the pilot at each start-up and the flight aborted if the engagement parameters are not met. In such a case it will be necessary to dismantle the clutch and inspect the linings for wear or contamination with oil. Needless to say that replacement is the only repair under these circumstances. A word of warning related to mercury clutches. Mercury forms an amalgam with many metals and once the process has started it cannot be reversed or stopped. Replacement of the affected component is the only remedy. Be very careful to avoid any mercury spillage and make sure that any that is spilt accidentally is collected immediately.

Main gearbox

Typically the main gearbox of a light helicopter contains two or three stages of speed reduction and provides the link between the rotor and the airframe that carries the whole weight of the helicopter (fig.10.12). It is a very important component and, like the main rotor, there is no form of emergency backup available if it should fail. For this reason the amount of work that is permitted outside an overhaul workshop is limited. Most of the routine work is limited to inspection and to servicing of the lubrication system. Nevertheless such inspection work is very important if incipient trouble is to be spotted at an early stage before it results in total failure. The main gearbox often, but not always, incorporates the freewheel unit and its operation is checked by the pilot during each run-up check prior to take-off. However it should also be checked during each inspection of the helicopter and this is easily done by turning the rotor in the normal direction of rotation. The rotor should be perfectly free to rotate and there should be no unusual noises when it is turned. If the helicopter has a manual clutch then this should be engaged to ensure a positive indication of the correct functioning of the freewheel. Where a centrifugal clutch is fitted the indication may not be so positive. Rotation of the rotor in the reverse direction should engage the freewheel and the drag of other parts of the transmission should be felt.

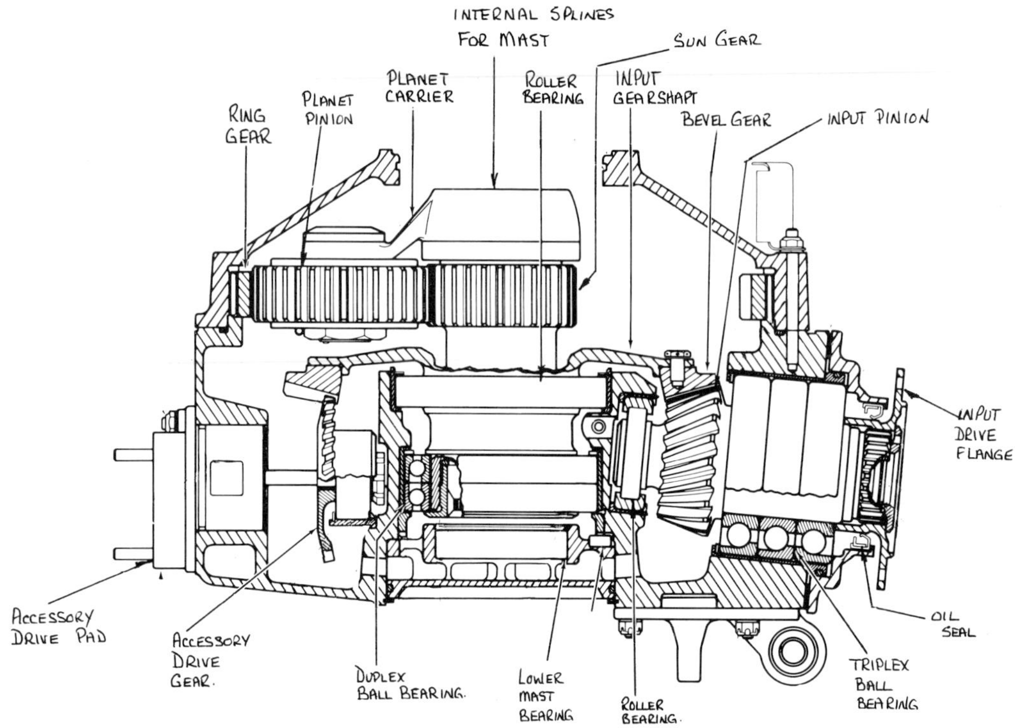

Fig. 10.12 Main gearbox.

The reduction gearing is usually epicyclic in form, there being one stage above the other. The planetary carrier of the upper stage is splined to transmit the drive to the mast. The lift loads of the helicopter are taken by a large thrust bearing in the mast-bearing housing that bolts to the top of the gearbox. The sun gear of the lower planetary assembly is driven by a large bevel gear which in turn is driven by a small bevel gear on the input shaft from the engine. The input shaft and its supporting bearings are mounted in a removable housing that is known as a quill, and thus this one is described as the input quill. There will be other drive quills that mate with the large bevel gear for driving the tail rotor and other accessories. In many gearboxes it is necessary to remove the drive quills from time to time in order to inspect the condition of the gear teeth, and when this is done great care must be taken to retain the large shims with their particular quill. These shims adjust the backlash and the pitch alignment of the bevel gears and must not be changed outside an overhaul shop where the necessary checks can be properly made. When a gear is removed the teeth should be inspected for any obvious wear, break-up of the surface and pitting caused by overloading or corrosion. In most cases either of these last two conditions will necessitate overhaul of the gearbox but there will be specific recommendations in the maintenance manual, often with photographs of acceptable and unacceptable wear patterns. The wear pattern is the polished portion of the flank of the gear tooth where the other tooth has been sliding over it under load. The shape and location of this wear pattern will vary with the form of the gear teeth and the backlash and pitch alignment of the gears.

Whilst the epicyclic gear has long been the industry standard there is a growing trend in gearbox design towards simple spur gear reduction trains. In some cases the trend was started by a military need to transport helicopters by air or hangar them on board ship and this is one way to reduce the overall height of the main gearbox.

Inspection of the main gearbox is not limited to the internal components. Since the casing of the gearbox carries the entire weight of the helicopter and is subject to all kinds of cyclic loading it must be carefully inspected for cracks during each routine maintenance check. Pay special attention to lugs and fittings that connect it to the airframe, bearing in mind that changes of section are often places of stress concentration. Installation of the gearbox into the airframe will follow the procedure written down by the manufacturer but must result in the installation being free of built-in stresses due to misalignment. Also do not forget to oil prime the gearbox on installation or the gears will be completely dry when the gearbox starts to rotate. In many cases it is recommended that a pint (0.5 l) of the correct oil is poured into the top of the gearbox casing just before the mast is fitted in place.

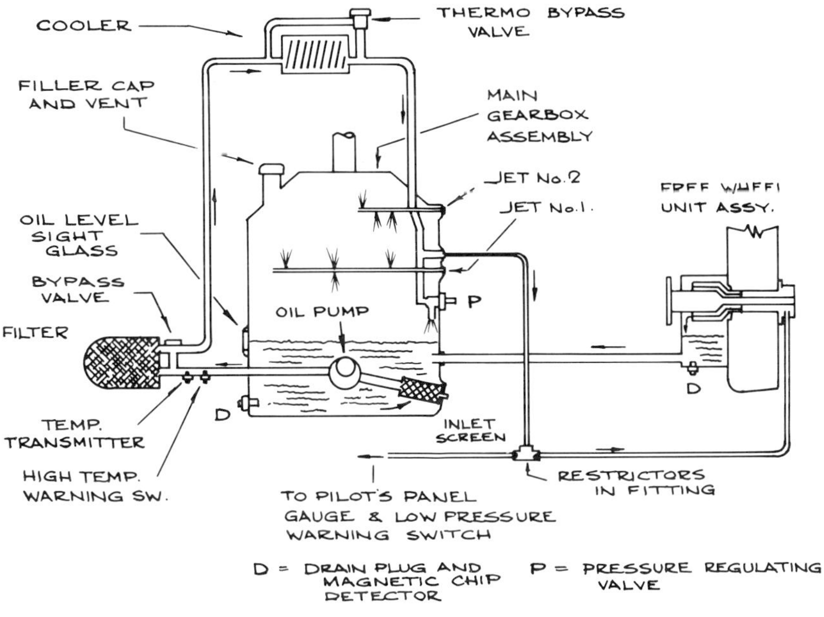

Fig. 10.13 Main gearbox oil system.

The main gearbox has its own pressure-fed lubrication system although in one case the oil is fed from the engine system (fig. 10.13). The oil is fed from an external tank, or from a sump integral with the gearbox, through a strainer to the pressure pump. The oil under pressure travels to a filter that is fitted with a bypass valve and so to the cooler and the lubricating jets. A pressure relief valve will be fitted in the system and this may be immediately following the pump or it may be at the end of the system, that is just before the jets. The scavenge oil returns to the sump by natural drainage, lubricating the various parts of the gear train on the way. Depending on the installation the oil cooler may have a separate cooling fan driven from the transmission, in many cases from the tail rotor shafting.

Periodic maintenance of the lubrication system consists of routine oil and filter changes and inspection of the chip detectors. After changing the oil and filter always carry out a ground run and check the installation for any leaks. The filter should be carefully cut open, where a paper element is used, and inspected for the presence of metallic particles and other deposits. There will always be a higher level of metallic particles when the gearbox is new or freshly overhauled but this should quickly diminish over the first 25 to 50 hours of operation. A sudden rise in the level after this period is cause for further investigation of the gearbox. A magnet should be used to separate the ferrous particles and the rest should be washed on to a clean laboratory filter paper using a little solvent. Brass and bronze can be identified by their colour but more positive identification can be made by using a series of simple chemical tests that are detailed in the table (fig. 10.14). Before any conclusions can be drawn it is important to know which materials are used in the specific gearbox being examined and where they are located. The shape of the particles should also be noted. Inevitably the judgment as to whether to leave the gearbox in service or not can only be one that is based on experience with that particular type and any recommendations given by the manufacturer.

Magnesium	A simple test is to try burning a few particles – it burns with a bright white flash.
Cadmium	Place the particles in a 10% solution of ammonium nitrate in water. If they are cadmium the particles will dissolve.
Tin	Tin particles can be identified by their low melting point. If the particles are dropped on to a clean tinned soldering iron at 260°C they will melt.
Aluminium	If a particle of aluminium is placed in 50% hydrochloric acid it will fizz with rapid emission of gas bubbles and gradually disintegrate to form a black residue. Silver and bronze do not noticeably react with hydrochloric acid.
Silver	When a silver particle is placed in nitric acid it reacts rather slowly, producing a whitish fog (silver nitrate) in the acid.
Bronze	Bronze or copper placed in nitric acid produces a bright green cloud.

Fig. 10.14 Material identification by chemical test.

Tail rotor drives

The drive to the tail rotor is taken from the main gearbox although in some turbine powered helicopters the drive is taken straight from the engine output gearbox. The drive is transmitted by shafting that is supported in hanger bearings along the tail boom to the tail rotor gearbox. This gearbox is used to turn the drive through 90° and often makes the final speed reduction. Where a cranked tail boom forms part of the helicopter design the drive changes direction by means of an intermediate gearbox or by use of a constant velocity universal joint. The gearboxes usually consist of a pair of bevel gears with their own splash lubrication system. Frequently the oil is taken to the bearings on the output shaft of the intermediate gearbox by means of an Archimedean screw machined into the output shaft.

The tail rotor drive shafts are supported on grease-lubricated ball bearings that are mounted in hangers attached to the structure of the tail boom. The bearings may be greased periodically or they may be packed for life but they need a little more attention than lubrication. On any occasion possible a check should be made to see whether grease is being thrown from the bearing as this is a sign that the bearing is overheating. A further check may be made by feeling each bearing soon after the helicopter has been shut down, for an ailing bearing will run at a higher temperature than the rest. Either of the above indications is good cause to replace the bearing. Some light helicopters use a one-piece tail rotor drive shaft and it is very important that the bearings are correctly aligned before the shaft is installed. This is achieved by stretching a thin wire between the end couplings and ensuring that the wire passes through the centre of each bearing. Larger helicopters always use multiple shafts that are connected to each other at the bearings by means of flexible couplings that allow for some misalignment and compensate for deflection of the airframe during flight. One common form of these couplings consists of a stack of thin sheet steel rings bolted to two adaptors at 90° to each other. The laminations should be inspected for signs of cracking, bowing or corrosion at the routine inspection periods. The stack of laminations should only be replaced as an assembly: individual laminations should not be replaced. The shaft itself is balanced as an assembly but, due to the nature of the loads it carries, it is sensitive to any surface damage such as scratches and nicks. Any of these could be the cause of a subsequent fatigue failure so the utmost care in handling and fitting these shafts is imperative.

CHAPTER 11
Cabin Air Systems

The comfort of the pilot is important if he is to function in an efficient manner and if his judgment is not to be impaired by distractions and discomforts. The passengers will be more relaxed and less liable to feeling unwell if they are comfortable irrespective of the state of the weather outside the aircraft. For these reasons light aircraft are fitted with systems that are able to provide a supply of fresh air in the right places and to heat it if required. Many of the more sophisticated light aircraft now provide a vapour cycle cooling unit as an optional extra but their application tends to be limited to the warmer climates.

Hardware

As with other systems in the aircraft the air systems have their own particular hardware which is fairly standard across the industry. The way in which the individual parts are used will depend upon the preferences of the designer and the size and sophistication of the aircraft in which they are installed (fig.11.1). All air systems must have a means of conducting the air from the inlets to the various outlets in the cabin and the most common means of doing this is by the use of wire-supported hoses. In this type of construction the fabric that constitutes the hose is supported in its circular shape by means of a continuous spirally wound piano wire on the inside. The fabric comes in two grades dependent upon the temperature at which it is to operate, the impregnation materials being silicone and neoprene. The former is used for the hose connections in the engine bay and those in the heater system and is red in colour. The latter has a smaller operating temperature range and is black in colour. Both come in a range of diameters and are cut to the length required.

The fabric of the hose may be readily cut using a sharp knife inserted between the wires. The wire is high carbon steel and must be cut with a pair of diagonal cutters that have hardened cutting edges since it will damage standard cutters and those found in standard pliers. Once the fabric and the wire have been cut, the last inch (25 mm) of the wire should be bent at 45° towards the open end of the hose. This end of the wire and the end of the outer reinforcing cord should be secured under the clamp once the hose has been fitted to the spigot of the connection. In this way the hose is properly supported and there is no danger of it collapsing.

This type of hose may be clamped to its fitting with any sort of clamp that is approved for the purpose but since the pressures involved are very low, lightweight clamps are usually employed. This type of clamp (the Tinnerman clamp) consists of a strip of spring steel one end of which is formed into a series of ratchet teeth and the other into a hook, or pawl, that can engage the ratchet. In use the clamp is correctly aligned in its proper location and squeezed together until the pawl engages in the

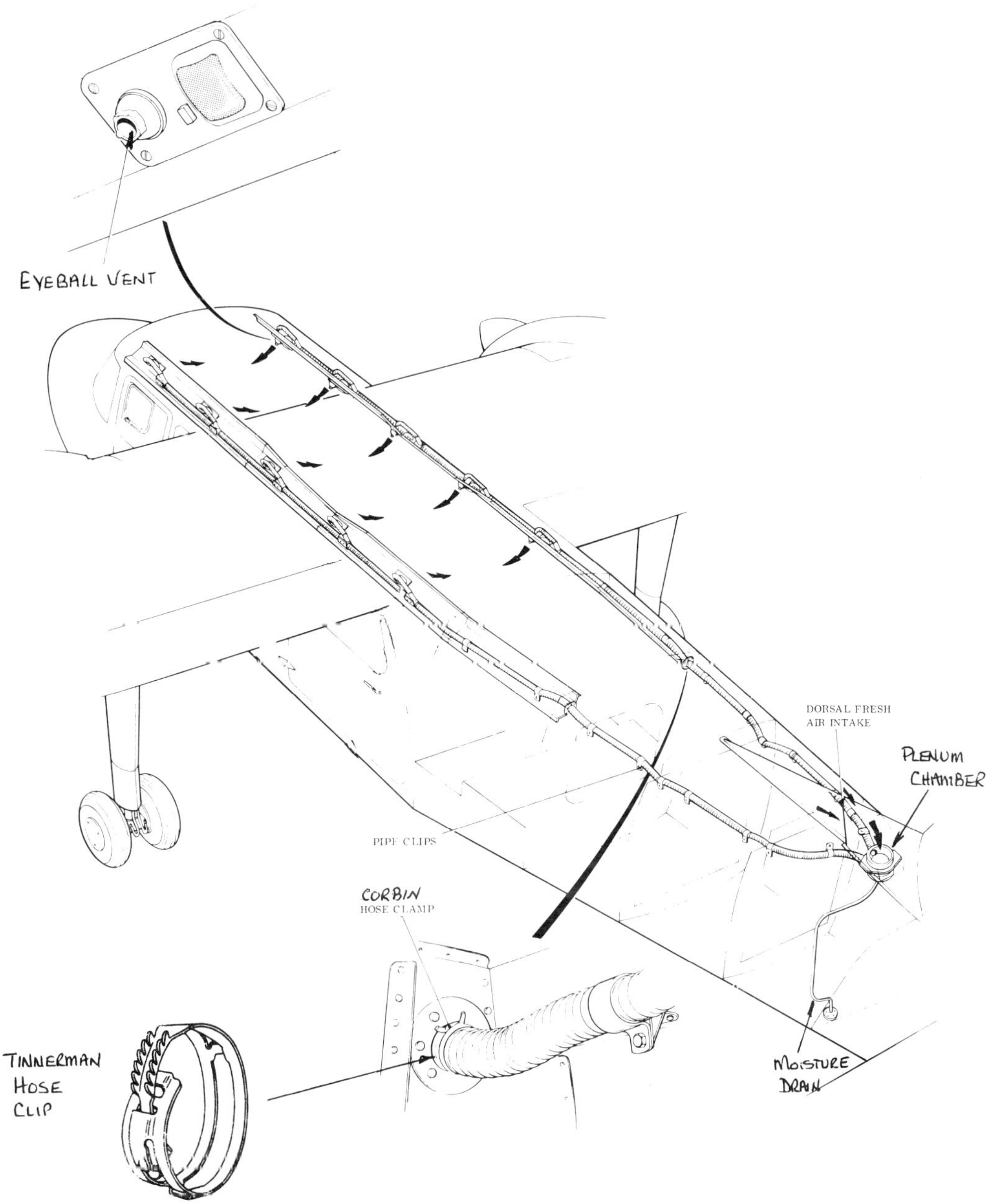

Fig. 11.1 Fresh air ventilation system.

ratchet teeth to maintain an appropriate tension. This is done most easily using a pair of water pump pliers (beak nose slip joint pliers). This type of tool opens readily to the appropriate size and the angled jaws permit the fitting of clips in awkward locations.

The wire-supported hose must be installed with care or it can be quite short lived and, in addition, do a lot of damage to other components. The routeing of the hose must be such that it is completely clear of other components and in many cases adjusting the length slightly will achieve this ideal. There are times when it is not possible to obtain ideal conditions and then the hose must be secured properly to avoid chafing. If the hose is allowed to chafe the wire will very quickly wear through the fabric and then through the component on which it is chafing. The hardened steel wire will make short work of electrical wiring, hoses and pipes and has been known to damage engine mount frames to the point where a welded repair was necessary. Tying with waxed string is rarely adequate since the thin section of the string tends to cut the hose. The solid plastic string ('spaghetti') is better provided it is tied in a figure-of eight method. Proprietary plastic ties (such as 'Tyraps') are quite good provided that the one round the hose is attached to the component with a second strap in order to prevent chafing. The most permanent and satisfactory method is to clamp the hose in a cushion (rubber covered) clamp and to attach this to the component in any suitable manner.

Cabin fresh air systems

In the typical light aircraft the fresh air system is very simple. Forward facing vents are located in the leading edges of the wings or in the leading edge of the vertical fin and the ram air ducted into the cabin through flexible ducts. The inlet aperture embodies a grille to prevent the entry of foreign matter. Since air entering at the forward speed of the aircraft tends to be somewhat noisy the inlet duct may embody a plastic filter similar to some domestic pan scourers. This slows the air down and acts as a fine filter. Should there be insufficient air in the cabin in a hot climate it is often necessary to remove this filter in order to increase the air flow. Another means of quietening the air flow is to duct the inlet air into a plenum chamber in which the air is allowed to expand and so decrease in velocity. The occupants of the cabin are able to adjust the airflow by means of variable opening vents, the most common being the 'eye-ball' unit which permits adjustment of flow and direction. The inlet duct may have a small hole drilled in the lower side at a low point in order to drain moisture from the duct. Occasionally drain holes are added at the low point of a hose.

Cabin heat systems

In single-engined aircraft it is standard practice for the air to be heated by a heat exchanger which forms part of the engine exhaust system (fig.11.2). Air is taken into the system from the engine baffles being the ram air above the cylinders. This air passes through the heat exchanger and into a plenum chamber in the cabin. From the plenum chamber ducts take the heated air to floor level outlets and to the windscreen where it is used for demisting purposes. The flow of the heated air is controlled by means of flap or butterfly valves in the system at appropriate places. The system is simple and quite reliable. Since the engine exhaust gases are used to heat the air the integrity of the heat exchanger is important if carbon monoxide is to be kept out of the cabin. As a result it is necessary to periodically pressure test the heat exchanger with it removed from the aircraft. The openings are blanked and the heat exchanger pressurised with air, usually to about 1½ p.s.i. (3 in. Hg). The entire assembly is then submerged in water when the presence of bubbles will reveal any leaks.

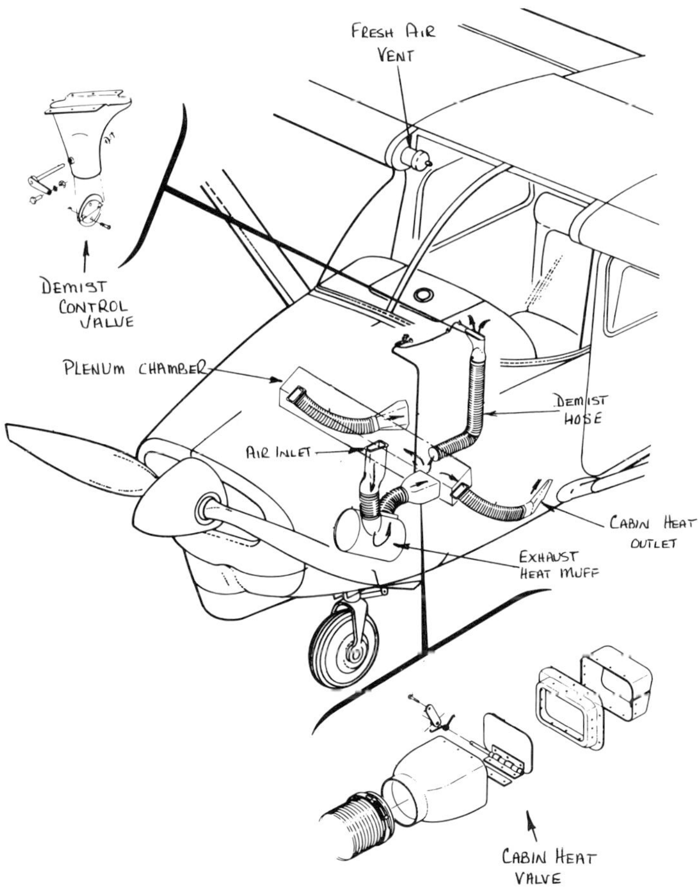

Fig. 11.2 Cabin heat system.

Combustion heaters

Twin-engined aeroplanes and some helicopters do not use exhaust heat exchangers but heat the cabin air by means of a fuel-fed heater, or combustion heater (fig.11.3). The combustion heater has the distinct advantage that heat is available without running the engines and, in addition, the output of the heater can be closely matched to the requirements of the cabin. Each heater unit consists of a combined combustion chamber and heat exchanger into which fuel is sprayed through a nozzle and ignited by a spark plug. Combustion air enters the combustion chamber tangentially and imparts a whirling action to the air, thus producing a whirling flame that is stable and sustains combustion under the most adverse conditions. The burning gases travel the length of the combustion tube, flow through the heat exchanger passages and out through the exhaust. Ventilating air passes through the heater between the jacket and the combustion tube outer surface and through an inner passage in the assembly before passing to the cabin.

The temperature of the air supplied to the cabin is sensed by a thermostat in the outlet duct which turns the fuel on and off as required by means of a solenoid valve incorporated in the nozzle. By cycling the heater on and off an even temperature is

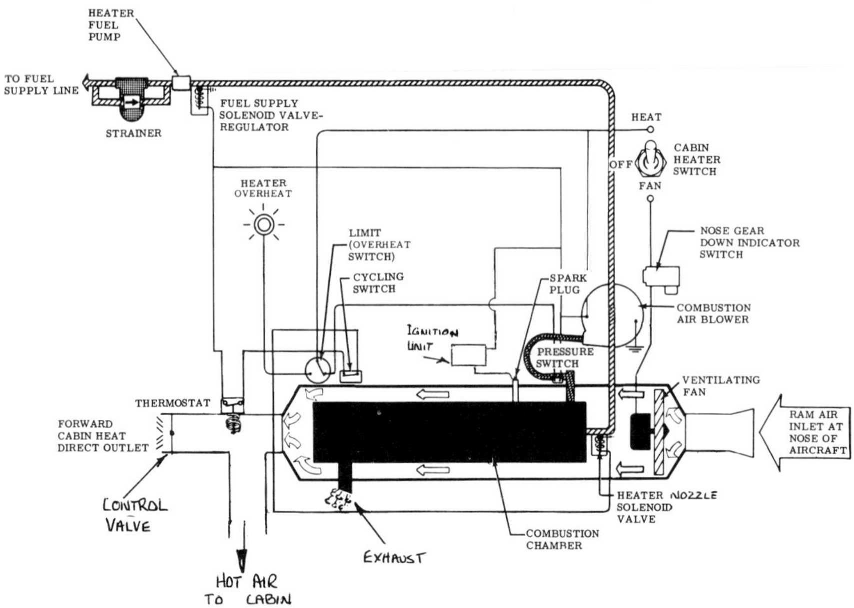

Fig. 11.3 Combustion heater.

maintained in the cabin. A continuous spark is provided at the spark plug from a transistorised ignition unit. Earlier model combustion heaters used an ignition system very similar to the coil ignition in a car, the contact breaker points being operated by the ventilating air blower. Still other models used an electro-mechanical vibrator to provide the interrupted D.C. to the ignition coil primary. Combustion air is provided by a separate tangential blower driven by an electric motor. A pressure or flow switch is included in the combustion air duct as a safety device and will cut off the electrical power from the nozzle solenoid valve, and so the fuel to the nozzle, in the event of a failure of the combustion air supply. Failure of the thermostat would cause the heater to overheat and protection is provided in the form of an overheat switch. This is set to a temperature considerably higher than the normal operating temperature of the heater and when it opens it cuts off all electrical power to the heater, although in one model heater the combustion air blower is left running to cool down the heater. The overheat switch is a lockout type of device in that it will not reset automatically and, in some aircraft, it illuminates a red light on the instrument panel to advise the pilot. The switch may be reset manually on the ground after the cause of the overheat has been investigated. A ventilating air blower is provided so that the heater may be used on the ground, but some aircraft include a microswitch on the landing gear that inhibits the blower in flight.

Combustion heater fault diagnosis

The precise procedure to be followed will not only vary with the type of heater installed but also the aircraft in which it is installed, and the manufacturer's service manual must be consulted for wiring details and specific checks. There are, however, procedures that are common to most types and these will be considered here.

The first thing to look for in fault diagnosis is to see whether both combustion and ventilating air blowers run. If neither run when the heater is on check the power supplies up to the heater terminal strip. If both blowers run but the heater does not ignite check the power at the input terminal of the ignition unit. If there is no power here check back through the circuit including combustion airflow or pressure switches for proper operation. If only the ventilating blower will run start at the overheat switch and work from there. If these checks are satisfactory check for electrical power to the fuel solenoid valve and loosen the fitting slightly to see if there is fuel under pressure. The solenoid coil can be checked for continuity using a multimeter. If the heater ignites but does not produce sufficient heat carry out a temperature check at a point about 12 - 18 in. (30 - 45 cm) downstream from the heater. With the thermostat set to the high position it should be possible to obtain a temperature in the region of 180–195°F (82–90°C). If this is not achieved it will be necessary to check the thermostat operation. If the fault is only evident at high airspeeds it is likely that there is a fault in the fuel system which will then require a proper flow check. If the heater is completely cold, but only at high airspeeds the fault may not only be in the fuel system but possibly the combustion air supply pressure may be low or the associated flow/pressure switch faulty. The cause of a heater that ignites but will not burn steadily can be almost anything that will cause erratic fuel flow or electrical power. It is worth checking the spark plug for fouling and proper gap. In ignition units that do not use a mechanically operated contact breaker the ignition system may be functionally checked by unscrewing the spark plug and earthing it to the heater case. If electrical power is applied to the ignition unit input terminal the plug should spark cleanly and continuously, but do keep your fingers out of the way. Spark plugs may be cleaned in the same way as piston engine spark plugs, except that an adaptor may be needed on the cleaning machine. Intermittent heater operation at low outside air temperatures may be due to water in the fuel freezing. If this is suspected the heater fuel system must be completely drained, including any filters and sumps. The aircraft fuel system should also be checked for signs of water in sumps, filters and tanks and appropriate action taken.

The manufacturer of the heater specifies an overhaul time in heater operating hours. Since these will not coincide with the aircraft hours it is worth fitting an hourmeter to the heater to avoid a significant financial penalty. There may also be a requirement to pressure test the heater chamber at certain periods and this is accomplished in the same manner as for an exhaust heat exchanger. The test pressure may be higher than for an exhaust heat exchanger (possibly 3–5 p.s.i., 6–10 in. Hg.) but this will be specified in the manufacturer's service manual.

Carbon monoxide

The presence of carbon monoxide in the cabin has already been mentioned as a possible danger. Carbon monoxide is an active poison that is colourless and odourless but in an aircraft it is often mixed with other gases that can be detected by smell. Concentrations of carbon monoxide (CO) exceeding 1 part in 20 000 parts of air (0.005%) are hazardous. A concentration of 1% will result in unconsciousness in a few minutes and death shortly afterwards. When the gas is taken into the lungs it combines with the haemoglobin in the blood producing a stable red compound; the haemoglobin cannot then combine with oxygen and oxygen starvation results. Oxygen starvation of the brain reduces a person's ability to reason and make decisions. Long exposure to low concentrations is as hazardous as short exposure to relatively high concentrations. The effects increase in severity with increase in altitude and are more severe for a heavy smoker.

Early symptoms of carbon monoxide poisoning are feelings of sluggishness, being too warm and tightness across the forehead. These early symptoms may be followed by more intense feelings such as headache, throbbing or pressure in the temples and ringing in the ears. These, in turn, may be followed by severe headache, general weakness, dizziness and gradual dimming of vision. Large accumulations of carbon monoxide in the body result in a loss of muscular power, vomiting, convulsions and coma. Finally there is a gradual weakening of the pulse, a slowing of the respiratory rate and death.

From the engineer's point of view it is important to ensure that carbon monoxide cannot enter the cabin. Regular inspection of heater heat exchangers has already been mentioned but sealing of engine firewalls is equally important. It is possible for exhaust gases to enter wheel wells and lift strut fairings and find their way into the cabin and these too must be properly sealed. Locations where engine controls, pipes and electrical cables pass through the structure must be suitably sealed. Exhaust tail pipes and combustion heater exhausts must be maintained in good condition and must extend the proper distance through the cowling or aircraft skin. Some aircraft manufacturers recommend exhaust inspections at 25-hour intervals due to the fact that cracks and holes can appear in a relatively short space of time.

Relatively simple tests may be made for the presence of carbon monoxide in the cabin. In one type of equipment a sample of air is drawn into a transparent tube containing material which changes colour according to the amount of carbon monoxide present. The actual amount is determined by comparing the colour in the tube with a colour standard provided with the instrument. Another type of test utilises a card with a circular spot of active material in the centre. The cover should be removed from the card and the spot exposed for 3 minutes. The spot will darken in the presence of carbon monoxide but humidity and other gases will also affect it. Such a card must not be left exposed for more than 30 days for then the indications are likely to be unreliable.

Oxygen systems

With the introduction of supercharged engines for light aircraft which permit the operation of the aircraft at higher altitudes it has been necessary to do something for the crew and passengers. As a result lightweight continuous flow oxygen systems have become available, even in many single-engined aircraft. The need for oxygen at altitude is explained by the reduction in atmospheric pressure with increase in altitude. In round figures the atmospheric pressure at 18 000 ft. is only half that at sea level and the transfer of oxygen across the membranes of the lung depends on the pressure of the oxygen. From sea level to 10 000 ft. the changes in oxygen saturation are not significant in that the saturation level falls from 97% to 92%. However the saturation level graph is not linear and above 10 000 ft. the saturation level of oxygen in the blood falls much more sharply. No flights should be made above 10 000 ft. without pressurisation or supplementary oxygen.

Before examining the oxygen system in detail let us sound a note of caution. From a personal safety point of view, it is important to remember that oil or grease will ignite violently in the presence of oxygen, and if the latter is under pressure an explosion may result. Therefore never handle any component of an oxygen system with greasy hands or whilst wearing greasy overalls. For the same reason never attempt to lubricate a stiff oxygen valve or coupling with an oil can. Should oil from another system accidentally spill on to part of the oxygen system then wash it off quickly. Needless to say smoking should not be allowed and all electrical power to the aircraft and power tools should be switched off before working on the oxygen system. Do not blow through oxygen lines using the workshop air supply for it inevitably contains a small amount of oil and moisture. Oxygen lines that are not in use should be blanked in order to prevent dirt or moisture ingress. Oxygen suitable for aircraft oxygen systems must be dry or the moisture will freeze at altitude and cut off the flow of oxygen when it is most needed. Breathing oxygen in the U.K. is supplied in black cylinders with the neck of the cylinder painted white. Where jointing compound is used it must be compatible with oxygen systems but many threaded unions are equally well sealed using PTFE tape. The jointing compound should be smeared on the first three threads of the male fittings only but should never be used on any flared fitting.

Storage cylinders

Military, transport jet and general aviation aircraft all use different systems that are dictated by their different operating environments and missions. Only the general aviation systems will be discussed here (fig.11.4). The systems used in light aircraft use a high-pressure storage bottle which is usually painted green for identification. Since the vast majority of these systems are of American manufacture hardware relevant to these systems will be considered. Two types of cylinder are in common use and are identified by their U.S. Government designations. The standard weight cylinder, identified by ICC or DOT-3AA 1800 stamped on the shoulder or neck of the cylinder, must be hydrostatically pressure tested to 3 000 p.s.i. (207 bars) every 5 years but there is no ultimate life limitation. The cylinders may continue in service until they fail pressure test or inspection. The lightweight cylinder, identified by ICC or DOT-3HT 1800, must be pressure tested to 3 000 p.s.i. (207 bars) every 3 years but also has an ultimate life of 24 years or 4 380 filling cycles, whichever occurs first. Due to the high stresses involved when these cylinders are filled to their working pressure of 1 800 p.s.i. (124 bars) quite small damage can be sufficent to scrap the cylinder. Typically for a DOT-3HT cylinder the maximum allowable depth of cuts and gouges and local corrosion is 0.005 in. (0.13 mm). For a DOT-3AA cylinder the maximum depth is about twice that specified for the lightweight cylinder. The existence of any defect outside these limits is reason for taking the cylinder out of service and sending it to an approved workshop capable of conducting the necessary inspections and tests.

Regulators and reducing valves

The earlier oxygen systems for light aircraft incorporated a manually controlled valve that was adjusted with every change of aircraft altitude. The automatic regulator, on the other hand, controls the oxygen pressure to the masks proportionately with the altitude and the only pilot control is an on-off valve. In both these systems oxygen is piped to the regulator at full cylinder pressure. Another system in common use avoids this high-pressure line by fitting a pressure-reducing valve directly to the storage cylinder. This valve incorporates an on-off valve and reduces cylinder pressure to approximately 70 p.s.i. (4.8 bars) at which pressure it is distributed to the

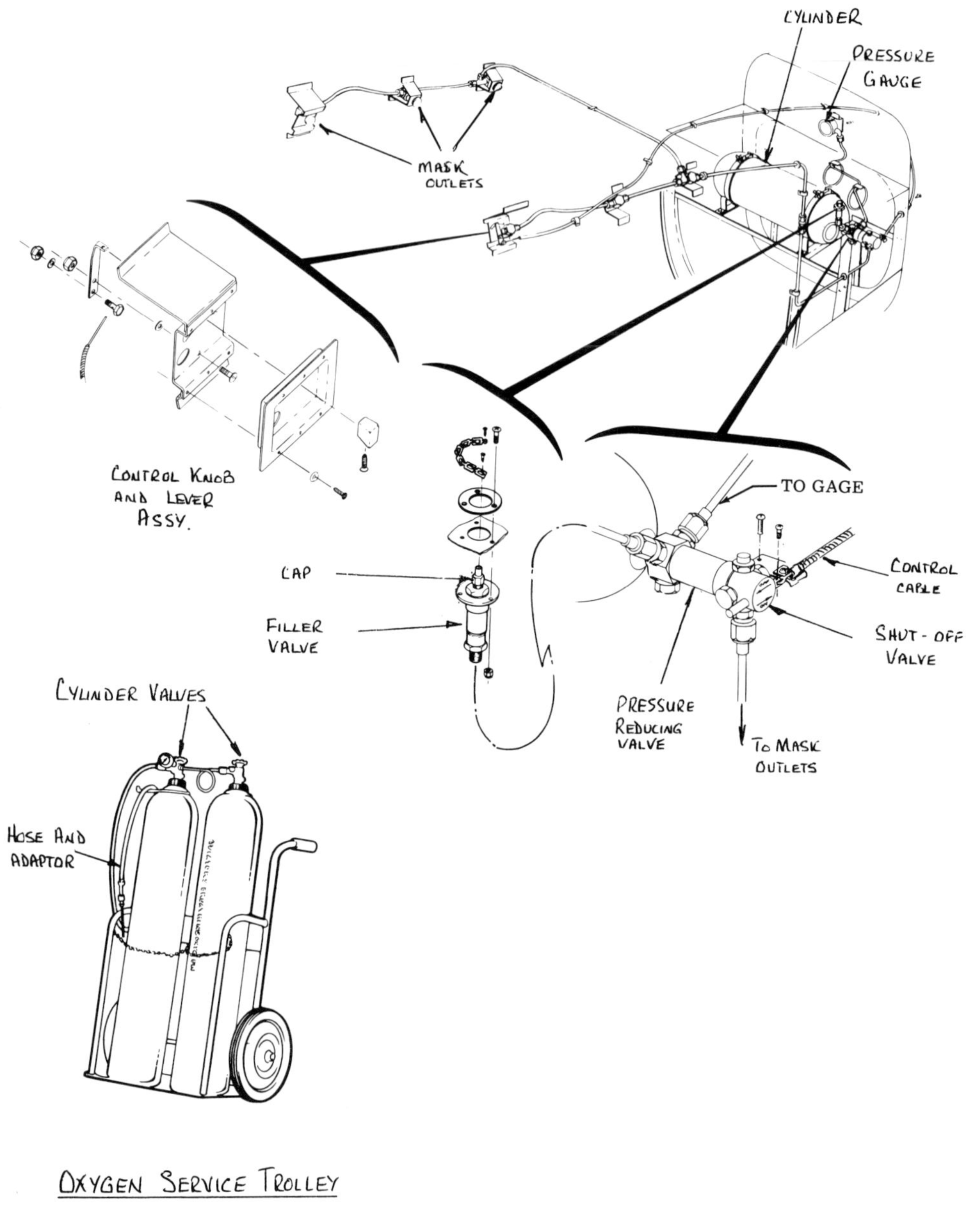

Fig. 11.4 Oxygen system.

cabin. Maintenance and repair is limited to changing a faulty component but in the case of a pressure reducer system the complete bottle and valve assembly should be repaired at a qualified overhaul shop.

Outlets and masks

Each outlet contains a spring-loaded valve that opens when a mask is plugged in and closes when it is unplugged. In order to limit the actual flow of oxygen a restrictor is fitted in the plug or line to each mask. By the use of different size restrictors the pilot is supplied with twice as much oxygen as each passenger to enable him to cope with the more demanding task of flying the aircraft. Where an automatic regulator is included in the system the restrictor size is fixed and is suitable for use up to the maximum altitude for which the aircraft has been certified. When a pressure-reducing valve controls the pressure to a fixed value it is necessary to use different size orifices for different flight levels and these are colour coded for identification purposes.

Colour Code	Operating Altitude
Red	8 000–14 000 ft
Orange	14 000–22 000 ft
Green	22 000–25 000 ft

The orange orifice could be used at all flight levels up to 22 000 ft. but would supply much more than is required at the lower levels. The use of the red orifice at the lower levels provides adequate flow and conserves the oxygen supply. An indicator is incorporated in the flexible tube to each mask to indicate whether or not the oxygen is flowing.

Charging the oxygen system

Refilling or charging the oxygen cylinder is straightforward but a little care should be taken. Do not attempt to refill a cylinder if any of the following conditions exist.

1 Contamination or corrosion of the cylinder, interconnections or filler valve.
2 Cylinder pressure test due.
3 Cylinder completely empty after the shut-off valve has been off for a length of time. (Return to overhaul facility for inspection.)

When filling the cylinder it will need to be filled to a pressure that is more or less than the nominal 1 800 p.s.i. dependent upon the ambient temperature. The temperature of the cylinder will rise by some 14°C (25°F) as it is filled.

The cylinder should be filled as quickly as possible and then allowed to cool to ambient temperature.

The following procedure may be used in conjunction with the table of filling pressures when charging a cylinder.

1 Connect the oxygen trolley to the filler valve. (An AN 805-3 nut with an AN800-3 cone union is required to adapt the trolley hose to most airframe mounted filler valves.)
2 Open the aircraft cylinder valve.

3 Slowly open the valve of the cylinder on the oxygen trolley having the lowest pressure and allow the pressures to equalise.
4 Close the oxygen trolley cylinder valve and repeat with the cylinder with the next highest pressure until the aircraft oxygen cylinder pressure gauge reads the appropriate pressure in the chart.
5 Shut off all valves, disconnect the trolley and cap the filler valve and trolley hose.
6 A leak check may be carried out on all connections using a mild soap solution.

Table of Oxygen Filling Pressures

Initial Temp.		Filling Pressure	Initial Temp.		Filling Pressure
°F	°C	p.s.i.	°F	°C	p.s.i.
0	−17.7	1 600	70	+21.1	1 925
10	−12.2	1 650	80	+26.6	1 950
20	−6.6	1 675	90	+32.2	2 000
30	−1.1	1 725	100	+37.7	2 050
40	+4.4	1 775	110	+43.3	2 100
50	+10.0	1 825	120	+48.8	2 150
60	+15.5	1 875	130	+54.4	2 200

Purging the oxygen system

Purging is required if the system has been dismantled and all the oxygen lost from the aircraft pipelines. This procedure should be carried out with the aircraft outdoors and is accomplished by plugging in all the masks with the oxygen system turned on. All doors, escape hatches and windows should be left open and the oxygen allowed to flow until it is completely odourless. The oxygen should then be turned off and the aircraft left undisturbed to allow the oxygen to dissipate from the cabin. It will be necessary to refill the cylinder on completion of the purging operation.

Functional testing of the oxygen system

The oxygen system may be functionally tested quite simply. An oxygen pressure gauge calibrated to 100 p.s.i. is required and should be fitted to a mask plug by means of a suitable hose capable of withstanding 100 p.s.i. pressure. Where the system uses a pressure-reducing valve the system is turned on and the gauge plugged in to a convenient mask outlet. It should read the correct pressure as specified in the Maintenance Manual (about 70-80 p.s.i.). Masks should be plugged into all the remaining outlets and the indicated pressure should remain within the allowed tolerance (typically ± 10 p.s.i.). The system should be turned off and allowed to bleed down to zero and the masks removed. The pressure on the test gauge must not rise above zero when observed for 1 minute. With a system utilising an automatic regulator the test is the same but the gauge pressure will be related to the atmospheric pressure at the time of the test.

CHAPTER 12
Ice and Fire Protection

Ice protection

Relatively few light aircraft are cleared for intentional flight in icing conditions but many, particularly twin-engined aircraft, are fitted with equipment that will enable them to deal with an unexpected build-up of ice on critical surfaces. The most popular system for this class of aircraft utilises inflatable rubber boots on the

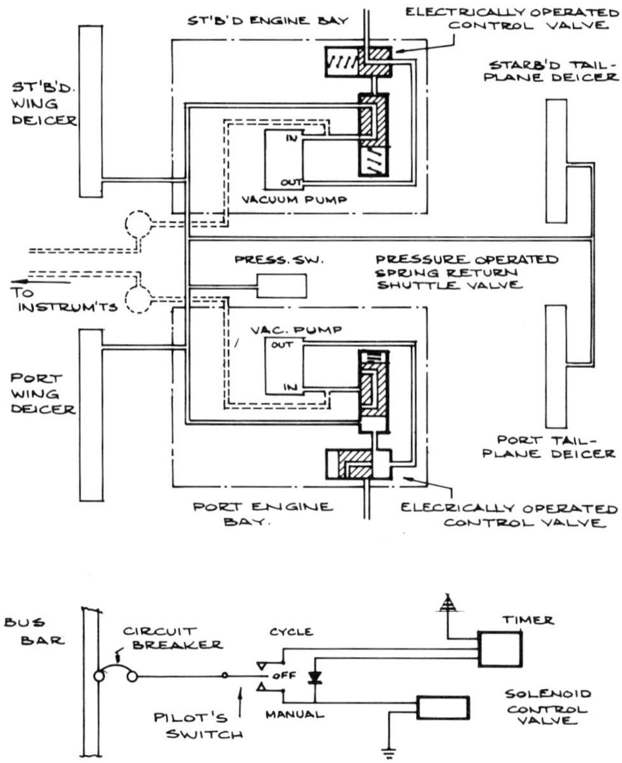

Fig. 12.1 Pneumatic de-ice system. Pneumatic circuit (top) and electrical circuit (bottom).

mainplane, tailplane and fin leading edges (fig.12.1). When ice has accumulated on the leading edges the boots are inflated and the ice broken away by the swelling of the boots. The cycle is repeated as necessary. Another system that enjoyed some popularity with a larger class of aircraft and is now finding some application with light aircraft is the fluid de-icing system. In this system an alcohol fluid is pumped to the leading edge areas and discharged through a porous metal plate that looks a little like the foil used on some electric razors. The alcohol is blown back over the aerofoil surface by the airflow and melts any accumulated ice. The windscreen also requires attention and this may vary from an alcohol spray or a small electrically heated panel in front of the pilot to a fully heated windshield. In addition all aircraft have the ability to direct the output of the cabin heater to the windshield for demisting and de-icing purposes. To complete the protection of the airframe the pitot head and the stall warning vane are heated electrically and an inspection light is built into the fuselage in such a way as to shine along the leading edge of the mainplane to help the pilot to identify ice build-up at night.

Pneumatic surface de-icing system

De-icer boots are made from soft, pliable rubberised fabric the outer ply of which is made from neoprene. Neoprene is used in this position in order to provide resistance to abrasion and deterioration and in order to provide an electrically conducting path to dissipate accumulated charges of static electricity. The whole boot is bonded to the leading edge of the aerofoil using a suitable adhesive, the edges are sealed and the job finished with a coat of conductive cement or paint to complete the electrical path to ground. Integral with the construction of the boot are longitudinal tubes which may be inflated by air pressure in order to break the ice loose from the surface (fig. 12.3).

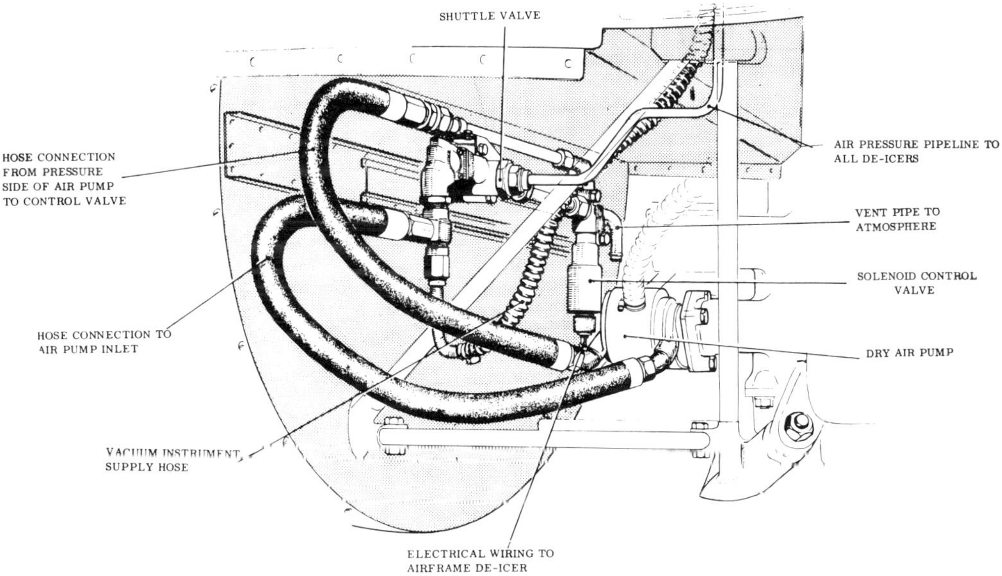

Fig. 12.2 Pressure and control components

With the boots deflated they are held to the contour of the leading edge by opening the tubes to the vacuum side of the vacuum pump that is used to drive the gyro instruments. The pressure to inflate the boots is supplied by the discharge side of the same pump. Where an oil lubricated pump is installed on the engine an oil separator is fitted between the pump pressure outlet and the de-ice pressure system. The pressure air is also passed through a filter before being admitted to the boots in order to prevent contamination and deterioration of the rubber. The routeing of the pressure and vacuum to the boots is controlled by a solenoid valve and a shuttle valve, although in some systems the solenoid valve combines both functions (fig. 12.2). Whilst the solenoid valve is de-energised the pressure air from the pump is

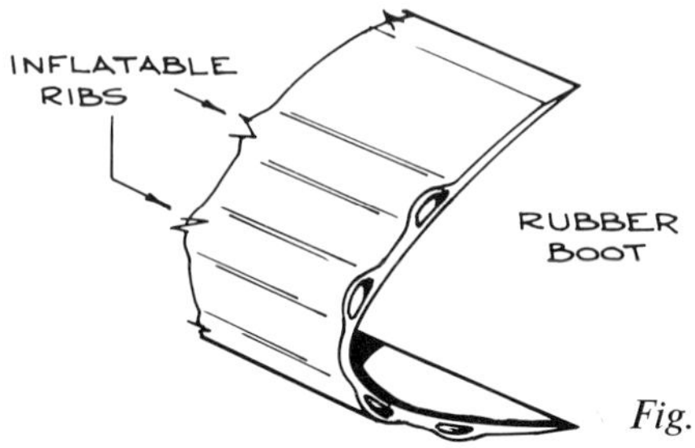

Fig. 12.3 De-icing boot cross section.

routed overboard through a breather line and no pressure is felt on the pressure port of the shuttle valve. When there is less than 1 p.s.i. (2 in. Hg) at this port the shuttle valve moves over and the boots are opened to the vacuum side of the pump which then holds them to the contour of the leading edge. When the pilot requires the de-icing function the solenoid valve is energised and air pressure is applied to the shuttle valve which moves over and admits the air pressure to the boots. A pressure switch senses this pressure and illuminates a light on the instrument panel. The pilot's control switch is biased to the OFF position. When it is selected to CYCLE and released an electromechanical or electronic timer controls a 6-second inflation cycle at the end of which the solenoid valve is de-energised. This allows the pressure to bleed overboard through the solenoid valve, thus causing the shuttle valve to move and admit vacuum to the boots. The cycle can be repeated by the pilot as often as is necessary by pressing the switch.

De-icer boot maintenance and repair

The service life of de-icer boots can be extended by proper care. The boots are particularly susceptible to scuffing damage caused by careless use of servicing steps and stands and refuelling hoses. Oil and fuel will attack the rubber and cause deterioration and the boots should be regularly cleaned with a mild soap and water solution to remove any contamination. The manufacturer's preparation known as ICEX should be applied sparingly to the boots in order to provide a smooth, polished film that will protect the surface of the boots and lower the adhesion of the ice to the boots.

In spite of the best care possible the boots will not last indefinitely and will need replacement when badly damaged or deteriorated. However minor damage can be quickly repaired with the boots in place on the aircraft. Damage in the tube area that is causing a leak may be repaired by using a fabric-reinforced patch cemented in place with the appropriate adhesive. The materials required are supplied in a kit designed for the job but the materials have a limited shelf life that should not be exceeded if reliable repairs are to be made. The procedure is similar to that used for patching an inner tube except that the patch must rolled down with a narrow wooden roller in order to ensure proper adhesion. The repair is completed by a thin coat of conductive cement.

The system may be readily checked by running the engines and operating the system. In the cockpit the advisory light should illuminate and the cycle should take a total of 6 seconds. An observer stationed outside the aircraft can observe that all the boots correctly inflate and deflate. This test will not detect small leakages that should really be repaired and a more accurate test may be conducted in the hangar where a compressed air supply is available. The pressure air hose is disconnected at the inlet port to the solenoid valve and a regulated source of compressed air is connected in its place. The inlet pressure should be about 20 p.s.i. (40 in. Hg) and may be conveniently supplied by a spray transformer or an engine differential compression tester equipped with suitable adaptors. With the pressure applied the system can be functioned normally to check the operation of the valves and pressure switch. The overboard breather line from the solenoid valve is now capped and the pressure applied and with the system isolated by means of a shut-off valve the leakage rate can be checked. Generally speaking it should not exceed 4 p.s.i. (8 in. Hg) per minute. Any leaks in the tubing connections or in the boots can be identified by brushing on a mild soap solution and looking for bubbles whilst the system is pressurised.

Fire protection systems

Single-engined light aircraft rarely boast any form of fire protection beyond a simple portable hand-held fire extinguisher. Many light twins are now offering fire protection systems for the engine bays as an optional extra. The systems offered fall naturally into two parts: those of fire detection and fire extinguishing, and each will be considered in turn.

Fire detection systems

In order to provide the pilot with a warning of a fire an electrical circuit is provided which is able to respond to temperature increase and provide a warning of this fact. The heart of the system is the detector, of which several are installed in areas of the engine bay which are subject to high temperature. Whilst the firewire system remains the firm favourite for large aircraft the increased complexity of control equipment tends to lead light aircraft manufacturers to install individual detectors which are thermally operated switches (fig. 12.4). As the temperature rises so the shell

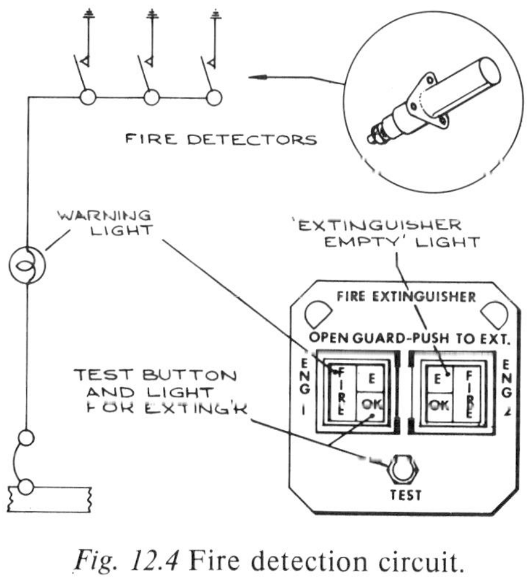

Fig. 12.4 Fire detection circuit.

of the detector expands and a spring-loaded switch is tripped, thus completing the circuit to a warning light. Some systems include an audible warning.

The system is simplicity itself and maintenance is also very simple. Since very few of these basic detection systems are provided with test circuits it is necessary to carry out periodic checks of the circuitry. Functional checks of the wiring may be carried out by shorting the terminal on the detector to ground with a suitable jumper lead. This test should be carried out for each detector in turn and the fire warning light should illuminate on each occasion. It should be noted that this test will not check the ground connection at each detector unless the jumper is touched to the metal shell of the detector. The operation of a thermal switch may be checked by gently playing the flame of a small propane blow torch over the end of the detector unit. The operating temperature may not be given in the aircraft service manual but it should be possible to judge an average speed of reaction when several detectors have been checked. The detectors in a circuit are in parallel with each other and the closing of any one switch will illuminate the warning light. A completely independent circuit is provided for each engine.

Fire extinguishing systems

The typical fixed fire extinguishing system for an engine bay stores the extinguishant under pressure in a spherical container (fig.12.5). The extinguishant is discharged

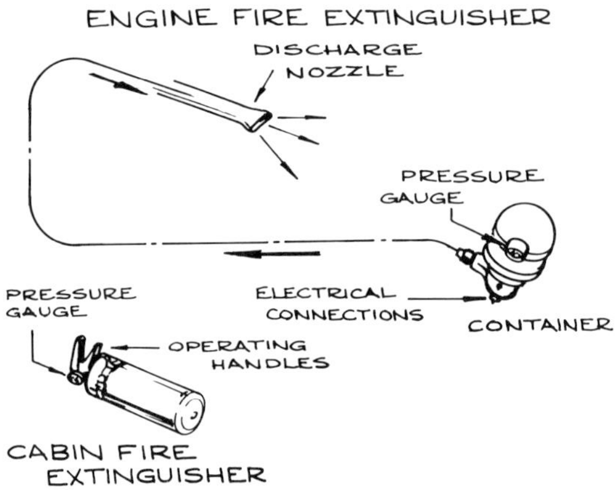

Fig. 12.5 Fire extinguishers.

into the engine bay when the pilot pushes the extinguisher firing button and electrically fires a pyrotechnic cartridge in the extinguisher bottle. This releases the extinguishant which is routed to the forward section of the engine where the natural airflow will distribute it throughout the engine nacelle.

Routine servicing consists of checking the fire extinguisher bottle pressure and electrical circuit checks. The spherical bottle is provided with a pressure gauge and the indicated pressure must be checked in relation to the ambient temperature. The information is provided in the service manual and may also be included on a transfer on the bottle. The variation of pressure with temperature may be quite wide, one example being from 300 p.s.i. (20.7 bars) at 4°C to 500 p.s.i. (34.5 bars) at 38°C. If the indicated pressure falls outside the tolerance allowed the bottle must be replaced.

Whenever working on the fire extinguisher system always make sure that all electrical power is disconnected from the aircraft.

The firing circuit of the fire extinguisher has a built-in test circuit to enable the continuity of the circuit to be checked without discharging the bottle. When the test button is pressed a very low current is passed through the circuit that is insufficient to fire the cartridge but which will illuminate a test light if the circuit is continuous. Should the test light not illuminate the wiring should be disconnected from the bottle by the plug and the appropriate pins shorted with a jumper wire. It the test light now operates satisfactorily the firing cartridge is faulty and must be replaced. Do not check the firing cartridge for electrical continuity using a continuity tester or a multimeter set to a resistance range - it is likely to fire the cartridge.

Hand-held fire extinguishers

All public transport aircraft are required to carry hand-held fire extinguishers, the number being dependent upon the size of the aircraft. The average light aircraft carries one extinguisher that is stowed within reach of the pilot.

The typical type of portable extinguisher is one that relies on a stored gas pressure to discharge the extinguishant. Squeezing the trigger or operating handle initially breaks a frangible seal inside the head and the subsequent discharge is controlled by operating the handle. Should this type of extinguisher be discharged the head must be removed, cleaned and fitted to a new cylinder. Another type is rechargeable and the pressure is retained by a valve operated by the discharge handle. The serviceability of these extinguishers is checked by one of two means, depending on the design. Some extinguishers are fitted with a pressure gauge, satisfactory pressure being indicated by a reading within the colour-coded arc on the dial. Another type indicates satisfactory pressure by means of an indicator disc on the base of the cylinder. If this disc can be pressed in with the thumb the pressure is low and the extinguisher is unserviceable. Such an extinguisher should be checked by the pilot before the first flight of the day to ensure that it is serviceable. Alternatively extinguishers are checked by weight against a known weight marked on the unit. Loss of weight indicates loss of extinguishant and the unit must be replaced. A weight check would be carried out less often than a pressure check, probably on a 6 to 12 month basis.

A variety of extinguishants are used in hand-held fire extinguishers but not all are suitable for all fires. Wet water is normal distilled water to which has been added an antifreeze and a wetting agent. It is suitable for fires in upholstery and furnishings but not for electrical or fluid fires. Dry powder extinguishers are effective against all classes of fire but when discharged in a confined space, such as a light aircraft cabin, can result in a severe problem regarding both visibility and breathing. BCF (bromochlorodifluoromethane) is also effective against all classes of fire and, unlike the dry powder extinguisher, requires no clearing up afterwards. The toxicity is low but it should be borne in mind that all these agents are designed to exclude oxygen in one way or another. As soon as the fire is extinguished it is necessary to thoroughly ventilate the compartment. If a dry powder extinguisher has been discharged it is very important to remove all traces of the powder since it can be corrosive to aluminium alloys if allowed to remain for an extended period.

CHAPTER 13
Aircraft Equipment

The title 'aircraft equipment' is intended to be a general one that brings together such items that do not conveniently fit under other headings. This does not mean that these items are any less important.

Seats

There is little doubt that the importance of seats is underrated by the vast majority of maintenance staff. Since the seat is the pilot's prime point of contact with the aircraft and provides the reaction point to most of his actions its failure can easily result in loss of control of the aircraft. In some cases the seat belts are attached to the seats which makes the seat structure even more important. Crew seats are often installed on tracks that allow them to slide fore and aft for adjustment purposes and the location of the seat on the rails is controlled by a locking device that inserts a bolt into a hole in the seat rail (fig.13.1). The function of this locking mechanism is critical

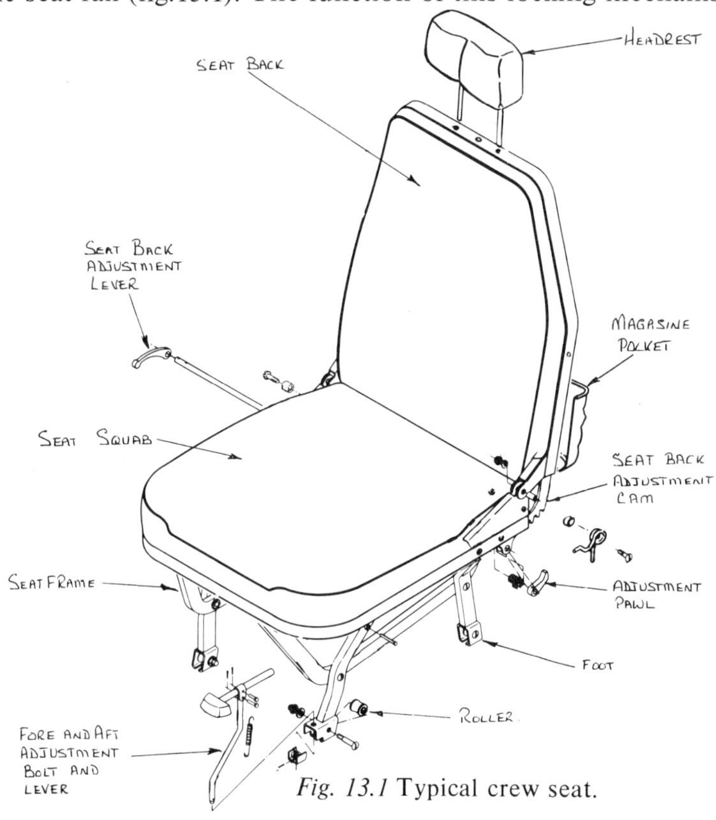

Fig. 13.1 Typical crew seat.

since failure will allow the seat to slide back to its full extent, taking the pilot out of reach of the controls. The bolt should freely enter each hole in the seat rail without any assistance from the pilot. The spring alone should be sufficient to fully engage the bolt. Where a bolt is used for each rail they must be so aligned that both bolts freely enter the locking holes together without any trace of binding. Seat travel stops must also be fitted to the rails to prevent the seat leaving the rails. The rails should also be carefully inspected since they are primary structural members. Some aircraft types are prone to the rails cracking through the locking holes and if this should occur the seat rails must be replaced.

Many seats in light aircraft have backs that will fold forward in order to facilitate access into the aircraft. Others possess seats that can be reclined. In both instances high stresses are generated at the points where the back attaches to the seat frame and this area should be inspected very carefully for cracks. Steel tube seat frames can usually be repaired by gas welding since no heat treatment is involved. Aluminium alloy frames are usually heat treated and although they may be welded using the heliarc process they are better replaced with a new item.

Elevating seats utilise a subframe to which the seat squab is fitted, the whole assembly being mounted on the mainframe. The vertical distance between the two can be varied by means of a screwjack and a series of bellcranks, thus providing adjustment to suit the pilot. Maintenance of this type of seat is similar to any other type except that the elevating mechanism requires periodic lubrication.

Seat belts and harnesses

Statistically the chances of you ever needing a seat belt or harness are small but when you do it must work first time for there can be no second chance. Because we need it on so few occasions there is a tendency for the humble seat belt to be neglected. Such neglect could be fatal – literally. And yet a seat belt or harness does not need a lot of maintenance. The assembly is only as strong as its attachment to the airframe so inspection should start there (fig.13.2). Make sure that the hardware listed in the parts catalogue has been used to attach the belts, that it is correctly locked and that the end fittings are free to swivel to allow the webbing to take up a natural line when under load. The webbing itself should be inspected for any signs of fraying or damaged stitching. Either of these defects will reduce the strength of the belt and should not be tolerated.

Where an inertia reel forms part of a harness installation give the shoulder strap a sudden jerk to make sure that it will lock properly. Inertia reels are usually set between 1g and 2g and so should lock readily. If the reel is fitted with a manual lock check that this operates properly. The webbing should wind back on to the reel freely and without creasing or fouling the sides of the reel housing. The operation of the buckle can be checked by sitting in the seat and tightening the seat belt and harness fully. Load may now be applied by arching your back and hips and then checking the operation of the buckle release. If the buckle shows any hesitation in releasing it should be replaced or repaired.

A full test, known as a proof load check, involves loading the belt by means of weights to one half of its rated strength. The rated strength is marked on belts of American manufacture but would have to be obtained from the manufacturer's literature in other cases. An American seat belt certified to TSO C-22 for one occupant has a rated strength of 1 500 lb (681 kg) but the webbing used in its

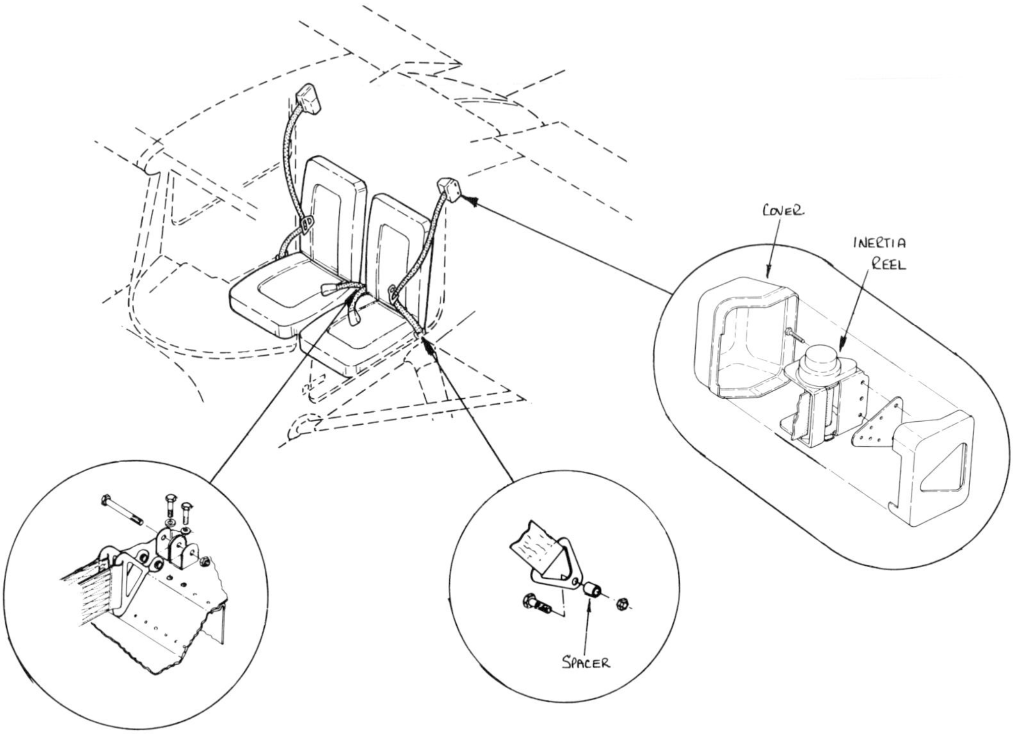

Fig. 13.2 Lap and diagonal seat belt installation.

construction may have a tensile strength of 5 000-6 000 lb (2 272-2 727 kg). The rated strength is for the assembly and takes into account the strength of the stitching and the geometry of the belt and applied loads. Once the appropriate load has been applied it should then be reduced to 125 lb (57 kg) and the buckle tested for satisfactory release. The pull required to operate the buckle release under these conditions must not exceed 45 lb (20 kg). Following the proof test the assembly should be carefully inspected for any signs of distress or damage to the stitches or fittings and the date of test should be marked on a suitable label sewn to the belt.

Seat belts do become dirty in use and may be washed in warm water using a mild soap and a nail brush. This should not be attempted in the aircraft and the belts should be allowed to dry naturally and thoroughly before refitting. Dry cleaning solvents should not be used to clean seat belts. Obviously care should be taken to keep the soap and water out of the release mechanism.

Soft furnishings

Carpets and upholstery used in aircraft are very similar to those used in the automotive industry. They differ from domestic materials in that the aircraft version has been treated to make it fire resistant. For this reason dry cleaning solvents should never be used unless the material is to be retreated. Generally speaking warm water and a mild soap may be used to clean carpets and upholstery provided note is taken of a simple word of warning. Wet carpets and light alloy floors do not go together. The carpet should be removed for cleaning and thoroughly dried before it is refitted or extensive corrosion of the structural floor could result.

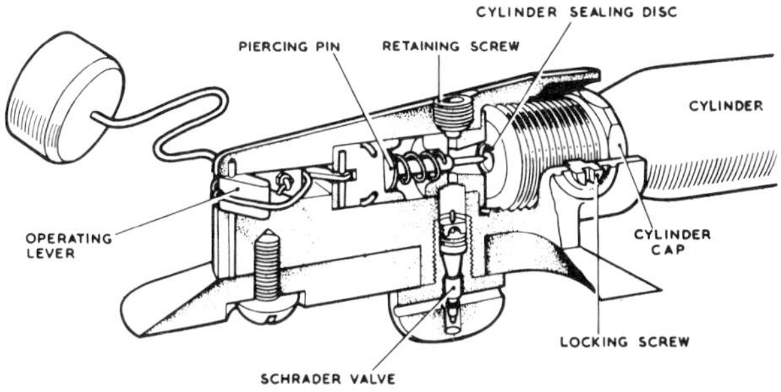

LIFEJACKET INFLATION HEAD

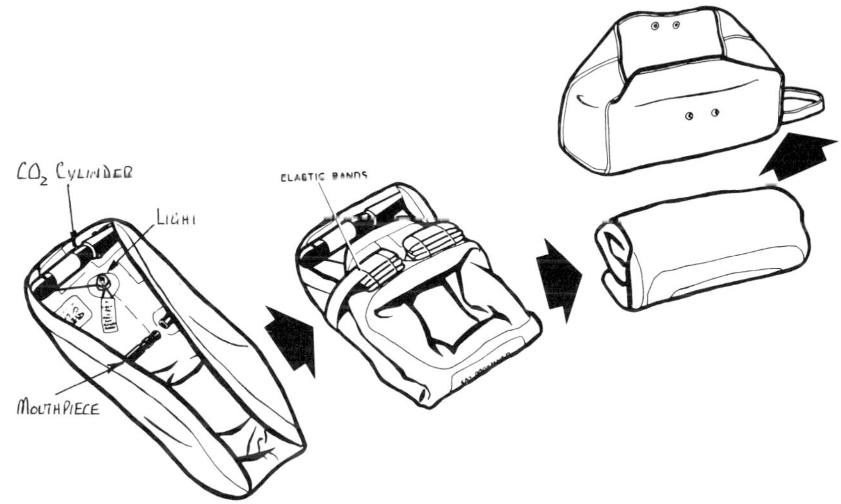

LIFEJACKET PACKING PROCEDURE

Fig. 13.3 Lifejacket servicing.

Behind the upholstery will be found soundproofing material, often in the form of mats of glass fibre insulating material. If this is renewed at any time each mat must be sealed into a polythene bag to prevent ingress of moisture. The glass fibre mat will absorb water by capillary action if it is not bagged. The plastic bag may easily be glued to the structure to retain the insulation in its proper place.

Lifejackets

Because of their construction lifejackets deteriorate with age and any misuse or improper storage will accelerate this deterioration. As part of every preflight inspection the lifejackets should be checked to ensure that they have not passed the life expiry date marked on the container or label.

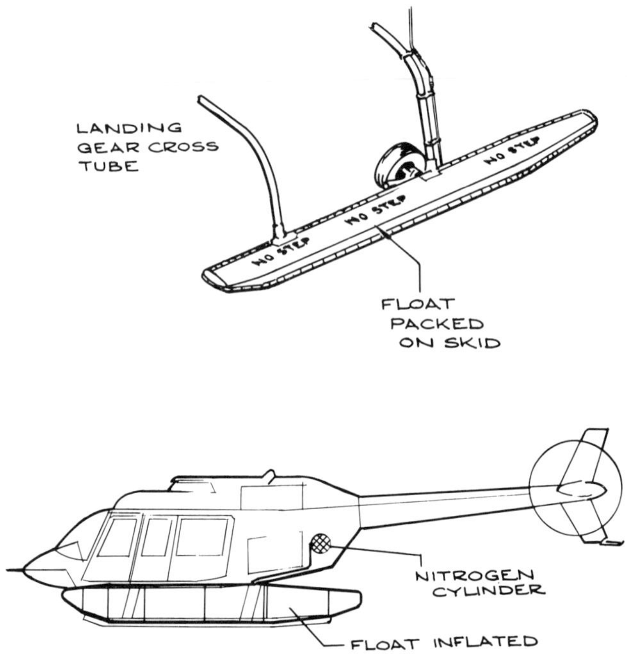

Fig. 13.4 Helicopter flotation gear.

Other checks required are simple enough but they do require a clean workshop. Cleanliness is very important when working with safety equipment. There are two main checks to be carried out to ensure serviceability - pressure test and CO_2 cylinder weight check. The CO_2 cylinder can be unscrewed from the operating head after the locking screw has been slackened (fig.13.3). The cylinder must be weighed on an accurate scale and the empty weight, stamped on the cylinder, subtracted from the total. This gives the weight of the gas charge which, in a typical example, must lie between 27 and 29 g. The actual figure will be given in the manufacturer's service manual.

The jacket may now be inflated using the mouth tube to check the functioning of the valve. Any major leaks should be immediately apparent and are cause for rejecting the lifejacket. Deterioration of the fabric is not easily seen, but any discoloured areas (other than those due to dirt), or sticky, stiff, shiny, wrinkled or crazed areas are signs that deterioration is taking place.

A full pressure test requires a source of low pressure air and a means of measuring low pressures, such as a water manometer. Using this equipment the lifejacket is inflated to a specified figure (about 1 p.s.i., or 28 in. of water) and the air supply isolated. The pressure loss due to leakage should not exceed about 5 in. of water in half an hour. The accessories such as whistles, lights, lanyards should all be checked for proper condition and operation. Refolding and packing the jacket is facilitated by sucking the air out with a vacuum cleaner. The jacket may then be folded and packed as in the illustration. The date of test or the new expiry date should be marked on the label before the jacket is returned to the aircraft.

Helicopter emergency flotation gear

To provide buoyancy for a helicopter that has to make an emergency landing on water, inflatable bags are provided in strategic areas of the aircraft (fig.13.4). These bags are packed in lace-up stowages on the outside of the helicopter or on the landing skids and can be inflated rapidly by high-pressure nitrogen or helium. The inflation can be either manually controlled by the pilot or, in some helicopters, the inflation is automatic being initiated by immersion switches fitted to the belly of the helicopter. Where the system is electrically controlled the compressed gas is released by firing pyrotechnic cartridges on each high-pressure storage cylinder in much the same way as for fire extinguisher bottles.

Inspection and maintenance of the gas system is similar to that required for a fire extinguisher system. The stored pressure is much higher, often as high as 3 000 p.s.i. (207 bars), but it will vary with temperature and must be checked on the pressure gauge provided on the gas cylinder. Periodically the bags are unpacked and inspected in a similar manner to a lifejacket, although the procedures and inflation pressures will differ due to the different size of the components. With the floats fully inflated (it may be necessary to place the helicopter on jacks) they should be lifted up to the position that they would occupy with the helicopter in the water and whilst they are in this position an inspection should be made to ensure that the fabric cannot chafe on any sharp edges, fasteners or other fittings on the helicopter. As with a lifejacket the float bags should be deflated for packing with the assistance of a vacuum cleaner. It may be necessary to remove the floats from the helicopter in order to pack them, but in any case make sure that they are clean since any grit or dirt can cause chafing and failure of a float bag on a subsequent inflation.

Bibliography

Title	*Author or Reference*	*Publisher*
Acceptable Methods, Techniques and Practices	AC 43-13	FAA
The Aeroplane Structure	A. C. Kermode	Pitman
Aircraft Air Conditioning Systems	EA-AAC-1	Aviation Maintenance Publishers
Aircraft Bonded Structure	EA-NMR	Aviation Maintenance Publishers
Aircraft Corrosion Control	EA-CC-1	Aviation Maintenance Publishers
Aircraft Electrical Systems	E. H. J. Pallett	Pitman
Aircraft Fabric Covering	EA-ADF	Aviation Maintenance Publishers
Aircraft Hydraulic Systems	EA-AH-1	Aviation Maintenance Publishers
Aircraft Instruments	E. H. J. Pallett	Pitman
Aircraft Oxygen Systems	EA-AOS-1	Aviation Maintenance Publishers
Aircraft Painting and Finishing	EA-AP-1	Aviation Maintenance Publishers
Aircraft Sheet Metal Construction and Repair	EA-SMF	Aviation Maintenance Publishers
Airframe and Powerplant Mechanics General Handbook	AC 65-9	FAA
Airframe and Powerplant Mechanics Airframe Handbook	AC 65-15	FAA
Air Navigation Order		HMSO
Air Navigation (General) Regulations		HMSO
Airworthiness Certification	CAP 396	CAA
Airworthiness Notices		CAA
Applied Science for the Aviation Technician	EA-AS	Aviation Maintenance Publishers
Automatic Flight Control	E. H. J. Pallett	Granada Publishing

Basic Science for Aerospace Vehicles	McKinley and Bent	McGraw-Hill
Civil Aircraft Inspection Procedures		CAA
Electricity and Electronics for Aerospace Vehicles	McKinley and Bent	McGraw-Hill
General Aviation Safety Information Leaflets		CAA
A Guide for Inspecting, Maintaining and Monitoring Aircraft Tyres	XAP-575	Goodyear Aviation Products
The Helicopter	J. Fay	David & Charles
Helicopter Flight Theory for Pilots and Mechanics	J. R. Montgomery	Sikorsky Aircraft
Light Aircraft Inspection	J. E. Heywood	T. & A. D. Poyser
Man at Altitude		Cessna Aircraft Corporation
Maintenance and Repair of Aerospace Vehicles	McKinley and Bent	McGraw-Hill
Mechanics of Flight	A. C. Kermode	Pitman

Aviation Maintenance Publishers, 211 South Fourth Street, Basin, WY 82410, U.S.A.

Civil Aviation Authority, Printing and Publication Services, 37 Gratton Road, Cheltenham, Glos. GL50 2BN.

FAA Publications are available from:
Aviation Maintenance Publishers, 211 South Fourth Street, Basin, WY 82410, U.S.A.,

Superintendent of Documents, U.S. Government Printing Office, Washington, D.C., 2042, U.S.A.

or

CSE Aviation Ltd., Airport Shop, Oxford Airport, Kidlington, Oxford OX5 1RA.

Data and Tables

Tapping and Clearance Drills for Screw Threads

	Size	Tapping Drill (mm)	Clearance Drill (mm)
B.A.	0	5.10	6.00
	1	4.50	5.40
	2	4.00	4.80
	3	3.45	4.20
	4	3.00	3.70
	5	2.65	3.30
	6	2.30	2.85
	7	2.05	2.55
	8	1.80	2.25
B.S.F.	1/4	5.30	6.50
	5/16	6.75	8.00
	3/8	8.25	9.80
	7/16	9.70	11.50
	1/2	11.10	13.10
B.S.P.	1/8	8.75	9.80
	1/4	11.80	13.50
	3/8	15.25	17.00
	1/2	19.00	21.25
A.N.F.	4	2.40	2.95
	6	2.95	3.70
	8	3.55	4.40
	10	4.10	5.00
	1/4	5.40	6.50
	5/16	6.90	8.00
	3/8	8.40	9.80
	7/16	9.90	11.50
	1/2	11.50	13.10
N.P.T.	1/8	8.30	10.50
	1/4	11.10	14.00
	3/8	14.25	17.70
	1/2	17.70	21.75

B.A. = British Association
B.S.F. = British Standard Fine
B.S.P. = British Standard Pipe
A.N.F. = American National Fine
N.P.T. = National Pipe Thread (American)

Standard Drill Sizes and

Decimal Equivalents

Standard Sizes			Standard Sizes			Standard Sizes			Standard Sizes			Standard Sizes			Standard Sizes		
in.	mm	Dec.	in.	mm	Dec.	in.	mm	Dec.	in.	mm	Dec.	in.	mm	Dec.	in.	mm	Dec.
	0.20	0.0079		1.65	0.0650		4.50	0.1772		8.40	0.3307	31/64		0.4844		18.50	0.7283
	0.22	0.0087		1.70	0.0669		4.60	0.1811		8.50	0.3346		12.40	0.4882	47/64		0.7344
	0.25	0.0099		1.75	0.0689		4.70	0.1850		8.60	0.3386		12.50	0.4921		18.75	0.7382
	0.28	0.0110		1.80	0.0709	3/16		0.1875		8.70	0.3425		12.60	0.4961		19.00	0.7480
	0.30	0.0118		1.85	0.0728		4.80	0.1890	11/32		0.3438	1/2	12.70	0.500	3/4		0.7500
	0.32	0.0126		1.90	0.0748		4.90	0.1929		8.80	0.3465		12.80	0.5039		19.25	0.7579
	0.35	0.0138		1.95	0.0768		5.00	0.1968		8.90	0.3504		12.90	0.5079	49/64		0.7656
	0.38	0.0150	5/64		0.0781		5.10	0.2008		9.00	0.3543		13.00	0.5118		19.50	0.7677
1/64		0.0156		2.00	0.0787	13/64		0.2031		9.10	0.3583	33/64		0.5156		19.75	0.7776
	0.40	0.0157		2.05	0.0807		5.20	0.2047	23/64		0.3594		13.10	0.5157	25/32		0.7812
	0.42	0.0165		2.10	0.0827		5.30	0.2087		9.20	0.3622		13.20	0.5197		20.00	0.7874
	0.45	0.0177		2.15	0.0846		5.40	0.2126		9.30	0.3661		13.30	0.5236	51/64		0.7969
	0.48	0.0189		2.20	0.0866		5.50	0.2165		9.40	0.3701		13.40	0.5276		20.25	0.7972
	0.50	0.0197		2.25	0.0886	7/32		0.2188		9.50	0.3740	17/32		0.5312		20.50	0.8071
	0.52	0.0205		2.30	0.0906		5.60	0.2205	3/8		0.3750		13.50	0.5315	13/16		0.8125
	0.55	0.0217		2.35	0.0925		5.70	0.2244		9.60	0.3780		13.60	0.5354		20.75	0.8169
	0.58	0.0228	3/32		0.0938		5.80	0.2283		9.70	0.3819		13.70	0.5394		21.00	0.8268
	0.60	0.0236		2.40	0.0945		5.90	0.2323		9.80	0.3858		13.80	0.5433	53/64		0.8281
	0.62	0.0244		2.45	0.0965	15/64		0.2344		9.90	0.3898	35/64		0.5469		21.25	0.8366
	0.65	0.0256		2.50	0.0984		6.00	0.2362	25/64		0.3906		13.90	0.5472	27/32		0.8438
	0.68	0.0268		2.55	0.1004		6.10	0.2402		10.00	0.3937		14.00	0.5512		21.50	0.8465
	0.70	0.0276		2.60	0.1024		6.20	0.2441		10.10	0.3976		14.25	0.5610		21.75	0.8563
	0.72	0.0283		2.65	0.1043		6.30	0.2480		10.20	0.4016	9/16		0.5625	55/64		0.8594
	0.75	0.0295		2.70	0.1063	1/4		0.2500		10.30	0.4055		14.50	0.5709		22.00	0.8661
	0.78	0.0307		2.75	0.1083		6.40	0.2520	13/32		0.4062	37/64		0.5781	7/8		0.8750
1/32		0.0312	7/64		0.1094		6.50	0.2559		10.40	0.4094		14.75	0.5807		22.25	0.8760
	0.80	0.0315		2.80	0.1102		6.60	0.2598		10.50	0.4134		15.00	0.5906		22.50	0.8858
	0.82	0.0323		2.85	0.1122		6.70	0.2638		10.60	0.4173	19/32		0.5938	57/64		0.8906
	0.85	0.0335		2.90	0.1142	17/64		0.2656		10.70	0.4213		15.25	0.6004		22.75	0.8957
	0.88	0.0346		2.95	0.1161		6.80	0.2677	27/64		0.4219	39/64		0.6094		23.00	0.9055
	0.90	0.0354		3.00	0.1181		6.90	0.2717		10.80	0.4252		15.50	0.6102	29/32		0.9062
	0.92	0.0362		3.10	0.1220		7.00	0.2756		10.90	0.4291		15.75	0.6201		23.25	0.9154
	0.95	0.0374	1/8		0.1250		7.10	0.2795		11.00	0.4331	5/8		0.6250	59/64		0.9219
	0.98	0.0386		3.20	0.1260	9/32		0.2812		11.10	0.4370		16.00	0.6299		23.50	0.9252
	1.00	0.0394		3.30	0.1299		7.20	0.2835	7/16		0.4375		16.25	0.6398		23.75	0.9350
	1.05	0.0413		3.40	0.1339		7.30	0.2874		11.20	0.4409	41/64		0.6406	15/16		0.9375
	1.10	0.0433		3.50	0.1378		7.40	0.2913		11.30	0.4449		16.50	0.6496		24.00	0.9449
	1.15	0.0453	9/64		0.1406		7.50	0.2953		11.40	0.4488	21/32		0.6562	61/64		0.9531
3/64		0.0469		3.60	0.1417	19/64		0.2969		11.50	0.4528		16.75	0.6594		24.25	0.9547
	1.20	0.0472		3.70	0.1457		7.60	0.2992	29/64		0.4531		17.00	0.6693		24.50	0.9646
	1.25	0.0492		3.80	0.1496		7.70	0.3032		11.60	0.4567	43/64		0.6719	31/32		0.9688
	1.30	0.0512		3.90	0.1535		7.80	0.3071		11.70	0.4606		17.25	0.6791		24.75	0.9744
	1.35	0.0532	5/32		0.1562		7.90	0.3110		11.80	0.4646	11/16		0.6875		25.00	0.9843
	1.40	0.0551		4.00	0.1575	5/16		0.3125		11.90	0.4685		17.50	0.6890	63/64		0.9844
	1.45	0.0571		4.10	0.1614		8.00	0.3150	15/32		0.4688		17.75	0.6988	1		1.0000
	1.50	0.0591		4.20	0.1654		8.10	0.3189		12.00	0.4724	45/64		0.7031	Standards rise		
	1.55	0.0610		4.30	0.1693		8.20	0.3228		12.10	0.4764		18.00	0.7087	in increments		
1/16		0.0625	11/64		0.1719		8.30	0.3268		12.20	0.4803		18.25	0.7185	of 1/64 in.		
	1.60	0.0630		4.40	0.1732	21/64		0.3281		12.30	0.4843	23/32		0.7188	and 0.50 mm		

Standard Drill Sizes Superseding Gauge and Letter Sizes

No.	Std. metric size
80	.35
79	.38
78	.40
77	.45
76	.50
75	.52
74	.58
73	.60
72	.65
71	.65
70	.70
69	.75
68	1/32
67	.82
66	.85
65	.90
64	.92
63	.95
62	.98
61	1.00
60	1.00
59	1.05
58	1.05
57	1.10
56	3/64
55	1.30
54	1.40
53	1.50
52	1.60
51	1.70
50	1.80

No.	Std. metric size
49	1.85
48	1.95
47	2.00
46	2.05
45	2.10
44	2.20
43	2.25
42	3/32
41	2.45
40	2.50
39	2.55
38	2.60
37	2.65
36	2.70
35	2.80
34	2.80
33	2.85
32	2.95
31	3.00
30	3.30
29	3.50
28	9/64
27	3.70
26	3.70
25	3.80
24	3.90
23	3.90
22	4.00
21	4.00
20	4.10
19	4.20

No.	Std. metric size
18	4.30
17	4.40
16	4.50
15	4.60
14	4.60
13	4.70
12	4.80
11	4.90
10	4.90
9	5.00
8	5.10
7	5.10
6	5.20
5	5.20
4	5.30
3	5.40
2	5.60
1	5.80

Letter	Std. metric size
A	13/64
B	6.00
C	6.10
D	6.20
E	1/4
F	6.50
G	6.60
H	17/64
I	6.90
J	7.00
K	9/32
L	7.40
M	7.50
N	7.70
O	8.00
P	8.20
Q	8.40
R	8.60
S	8.80
T	9.10
U	9.30
V	3/8
W	9.80
X	10.10
Y	10.30
Z	10.50

Common S.I./Metric Prefixes

Factor	Prefix	Symbol
1 000 000 = 10^6	mega	M
1 000 = 10^3	kilo	k
100 = 10^2	hecto	h
10 = 10^1	deca	da
0.1 = 10^{-1}	deci	d
0.01 = 10^{-2}	centi	c
0.001 = 10^{-3}	milli	m
0.000 001 = 10^{-6}	micro	μ

Standard Fluid Weights
(for weight and balance purposes)

Fluid	Weight					
	1 U.K. gall		1 U.S. gall		1 litre	
	lb	kg	lb	kg	lb	kg
Avgas	7.2	3.27	6.0	2.73	1.58	0.72
Avtur	8.1	3.68	6.75	3.07	1.78	0.81
Lubricating Oil	9.0	4.09	7.5	3.41	1.98	0.9
Water	10.0	4.55	8.3	3.77	2.2	1.0

Conversion Factors

	Multiply → To Obtain ←	By → By ←	To Obtain → Divide ←
LINEAR	inch	25.4	mm
	feet	0.3048	m
	yard	0.9144	m
	statute miles	0.8684	nautical miles
	statute miles	1.6092	km
	nautical miles	1.8529	km
	mil ('thou')	0.001	inch
AREA	sq. in	0.00695 ($^1/_{144}$)	sq. ft
	sq. in	6.4516	sq. cm
	sq. ft	0.1111 ($^1/_9$)	sq. yd
	sq. ft	0.0929	sq. m
VOLUME	cu. in	0.000578 ($^1/_{1728}$)	cu. ft
	cu. in	16.3871	cu. cm
	cu. in	1.6387×10^{-5}	cu. m
	U.K. gallon	1.2	U.S. gallon
	U.K. gallon	4.5454	litre
	U.S. gallon	3.7854	litre
	U.K. fl. oz	0.0625	U.K. pint
	U.S. fl. oz	0.05	U.S. pint
	U.K. gallon	0.004546	cu. m
VELOCITY	ft./min	0.00508	m/sec
	m.p.h.	0.8684	knots
	m.p.h.	1.6092	km/hr
	knots	1.8529	km/hr
MASS	pound	0.45359	kg
	U.K. ton (2 240 lb)	1016.736	kg
	U.S. ton (2 000 lb)	907.18	kg
	U.K. ton	1.0167	tonne
FORCE	pound	4.44822	Newton
	kg	9.80665	Newton
PRESSURE/STRESS	atmosphere	14.696	p.s.i.
	atmosphere	1013.3	m.b.
	atmosphere	29.9212	in. Hg
	p.s.i.	0.069	bar
	p.s.i.	2.036	in. Hg
	p.s.i.	27.6797	in. H_2O
	p.s.i.	6 894.76	Pascal (N/sq.m)
	mb	0.02953	in. Hg
	in. Hg	0.3	Pieze
	p.s.f.	4.784	daN/sq. m
POWER	H.P.	745.7	Watts
	metric H.P.	0.986318	H.P.
TORQUE	lb. ft	0.138255	kg-m
	lb. in	0.0113	metre-daN

Temperature Conversion

Numbers in the centre column (between those marked C and F) refer to temperature, either Centigrade or Fahrenheit, which it is desired to convert into the other scale. To convert from Fahrenheit to Centigrade find equivalent temperature in left-hand column marked C; in like manner find equivalent temperature in right-hand column when converting from Centigrade to Fahrenheit. Example: 50F. is 10C; 50C. is 122F.

C		F	C		F	C		F	C		F	C		F
-40.0	**-40**	-40.0	-1.1	**30**	86.0	26.7	**80**	176.0	71.1	**160**	320.0	137.8	**280**	536.0
-38.9	**-38**	-36.4	-0.6	**31**	87.8	27.2	**81**	177.8	72.2	**162**	323.6	140.6	**285**	545.0
-37.8	**-36**	-32.8	0.0	**32**	89.6	27.8	**82**	179.6	73.3	**164**	327.2	143.3	**290**	554.0
-36.7	**-34**	-29.2	+0.6	**33**	91.4	28.3	**83**	181.4	74.4	**166**	330.8	146.1	**295**	563.0
-35.6	**-32**	-25.6	1.1	**34**	93.2	28.9	**84**	183.2	75.6	**168**	334.4	148.9	**300**	572.0
-34.4	**-30**	-22.0	1.7	**35**	95.0	29.4	**85**	185.0	76.7	**170**	338.0	151.7	**305**	581.0
-33.3	**-28**	-18.4	2.2	**36**	96.8	30.0	**86**	186.8	77.8	**172**	341.6	154.4	**310**	590.0
-32.2	**-26**	-14.8	2.8	**37**	98.6	30.6	**87**	188.6	78.9	**174**	345.2	157.2	**315**	599.0
-31.1	**-24**	-11.2	3.3	**38**	100.4	31.1	**88**	190.4	80.0	**176**	348.8	160.0	**320**	608.0
-30.0	**-22**	-7.6	3.9	**39**	102.2	31.7	**89**	192.2	81.1	**178**	352.4	162.8	**325**	617.0
-28.9	**-20**	-4.0	4.4	**40**	104.0	32.2	**90**	194.0	82.2	**180**	356.0	165.6	**330**	626.0
-27.8	**-18**	-0.4	5.0	**41**	105.8	32.8	**91**	195.8	83.3	**182**	359.6	168.3	**335**	635.0
-26.7	**-16**	3.2	5.6	**42**	107.6	33.3	**92**	197.6	84.4	**184**	363.2	171.1	**340**	644.0
-25.6	**-14**	6.8	6.1	**43**	109.4	33.9	**93**	199.4	85.6	**186**	366.8	173.9	**345**	653.0
-24.4	**-12**	10.4	6.7	**44**	111.2	34.4	**94**	201.2	86.7	**188**	370.4	176.7	**350**	662.0
-23.3	**-10**	14.0	7.2	**45**	113.0	35.0	**95**	203.0	87.8	**190**	374.0	179.4	**355**	671.0
-22.2	**-8**	17.6	7.8	**46**	114.8	35.6	**96**	204.8	88.9	**192**	377.6	182.2	**360**	680.0
-21.1	**-6**	21.2	8.3	**47**	116.6	36.1	**97**	206.6	90.0	**194**	381.2	185.0	**365**	689.9
-20.0	**-4**	24.8	8.9	**48**	118.4	36.7	**98**	208.4	91.1	**196**	384.8	188.2	**370**	698.0
-18.9	**-2**	28.4	9.4	**49**	120.2	37.2	**99**	210.2	92.2	**198**	388.4	190.6	**375**	707.0
-17.8	**0**	32.0	10.0	**50**	122.0	37.8	**100**	212.0	93.3	**200**	392.0	193.3	**380**	716.0
-17.2	**1**	33.8	10.6	**51**	123.8	38.9	**102**	215.6	94.4	**202**	395.6	196.1	**385**	725.0
-16.7	**2**	35.6	11.1	**52**	125.6	40.0	**104**	219.2	95.6	**204**	399.2	198.9	**390**	734.0
-16.1	**3**	37.4	11.7	**53**	127.4	41.1	**106**	222.8	96.7	**206**	402.8	201.7	**395**	743.0
-15.6	**4**	39.2	12.2	**54**	129.2	42.2	**108**	226.4	97.8	**208**	406.4	204.4	**400**	752.0
-15.0	**5**	41.0	12.8	**55**	131.0	43.3	**110**	230.0	98.9	**210**	410.0	210.0	**410**	770.0
-14.4	**6**	42.8	13.3	**56**	132.8	44.4	**112**	233.6	100.0	**212**	413.6	215.6	**420**	788.0
-13.9	**7**	44.6	13.9	**57**	134.6	45.6	**114**	237.2	101.1	**214**	417.2	221.1	**430**	806.0
-13.3	**8**	46.4	14.4	**58**	136.4	46.7	**116**	240.8	102.2	**216**	420.8	226.7	**440**	824.0
-12.8	**9**	48.2	15.0	**59**	138.2	47.8	**118**	244.4	103.3	**218**	424.4	232.2	**450**	842.0
-12.2	**10**	50.0	15.6	**60**	140.0	48.9	**120**	248.0	104.4	**220**	428.0	237.8	**460**	860.0
-11.7	**11**	51.8	16.1	**61**	141.8	50.0	**122**	251.6	105.6	**222**	431.6	243.3	**470**	878.8
-11.1	**12**	53.6	16.7	**62**	143.6	51.1	**124**	255.2	106.7	**224**	435.2	248.9	**480**	896.0
-10.6	**13**	55.4	17.2	**63**	145.4	52.2	**126**	258.8	107.8	**226**	438.8	254.4	**490**	914.0
-10.0	**14**	57.2	17.8	**64**	147.2	53.3	**128**	262.4	108.9	**228**	442.4	260.0	**500**	932.0
-9.4	**15**	59.0	18.3	**65**	149.0	54.4	**130**	266.0	110.0	**230**	446.0	265.6	**510**	950.0
-8.9	**16**	60.8	18.9	**66**	150.8	55.6	**132**	269.6	111.1	**232**	449.6	271.1	**520**	968.0
-8.3	**17**	62.6	19.4	**67**	152.6	56.7	**134**	273.2	112.2	**234**	453.2	276.7	**530**	986.0
-7.8	**18**	64.4	20.0	**68**	154.4	57.8	**136**	276.8	113.3	**236**	456.8	282.3	**540**	1004.0
-7.2	**19**	66.2	20.6	**69**	156.2	58.9	**138**	280.4	114.4	**238**	460.4	287.8	**550**	1022.0
-6.7	**20**	68.0	21.1	**70**	158.0	60.0	**140**	284.0	115.6	**240**	464.0			
-6.1	**21**	69.8	21.7	**71**	159.8	61.1	**142**	287.6	116.7	**242**	467.6			
-5.6	**22**	71.6	22.2	**72**	161.6	62.2	**144**	291.2	117.8	**244**	471.2			
-5.0	**23**	73.4	22.8	**73**	163.4	63.3	**146**	294.8	118.3	**246**	473.0			
-4.4	**24**	75.2	23.3	**74**	165.2	64.4	**148**	298.4	121.1	**250**	482.0			
-3.9	**25**	77.0	23.9	**75**	167.0	65.6	**150**	302.0	123.9	**255**	491.0			
-3.3	**26**	78.8	24.4	**76**	168.8	66.7	**152**	305.6	126.7	**260**	500.0			
-2.8	**27**	80.6	25.0	**77**	170.6	67.8	**154**	309.2	129.4	**265**	509.0			
-2.2	**28**	82.4	25.6	**78**	172.4	68.9	**156**	312.8	132.2	**270**	518.0			
-1.7	**29**	84.2	26.1	**79**	174.2	70.0	**158**	316.4	135.0	**275**	527.0			

Pilot Maintenance

In the U.K. the Air Navigation Order 1980 Article 10(2) permits a licensed pilot who is the owner or operator of an aircraft not exceeding 2 730 kg maximum weight to *personally* make repairs on that aircraft. However this privilege does not apply if the aircraft is being operated for the purpose of public transport as defined in Article 93(6) of the same Order. The Order requires that details of the work carried out must be entered in the aircraft log book and the entry must be signed and dated. Since it also requires the use of approved spare parts the source and Approved Certificate number (or its equivalent) should be quoted for any replacement parts or materials used.

The above Article does not permit a pilot to carry out any type of repair or replacement and a list of permitted work items is given in the Air Navigation (General) Regulations 1981, Regulation 16. The list is as follows:

1. Replacement of landing gear tyres, landing skids or skid shoes.
2. Replacement of elastic shock absorber cord units on landing gear where special tools are not required.
3. Replacement of defective safety wiring or split pins excluding those in engine, transmission, flight control and rotor systems.
4. Patch-repairs to fabric not requiring rib stitching or the removal of structural parts or control surfaces, if the repairs do not cover up structural damage and do not include repairs to rotor blades.
5. Repairs to upholstery and decorative furnishing of the cabin or cockpit interior when repair does not require dismantling of any structure or operating system or interfere with an operating system or affect the structure of the aircraft.
6. Repairs, not requiring welding, to fairings, non-structural cover plates and cowlings.
7. Replacement of side windows where that work does not interfere with the structure or with any operating system.
8. Replacement of safety belts or safety harness.
9. Replacement of seats or seat parts not involving dismantling of any structure or of any operating system.
10. Replacement of bulbs, reflectors, glasses, lenses or lights.
11. Replacement of any cowling not requiring removal of the propeller, rotors or disconnection of engine or flight controls.
12. Replacement of unserviceable sparking plugs.
13. Replacement of batteries.
14. Replacement of wings and tail surfaces and controls, the attachments of which are designed to provide for assembly immediately before each flight and dismantling after each flight.

15. Replacement of main rotor blades that are designed for removal where special tools are not required.
16. Replacement of generator and fan belts designed for removal where special tools are not required.
17. Replacement of VHF communications equipment, being equipment which is not combined with navigation equipment.

In addition to the repairs shown above a licensed pilot is authorised as a signatory for the 50 hour check for an aeroplane, but not helicopter, with a Certificate of Airworthiness in the Private Category maintained to the Light Aircraft Maintenance Schedule (LAMS). As before the pilot must be the owner or operator of the aircraft, must make the necessary entries in the logbooks, must sign and date those entries and add the number of his licence. More details are given in the actual Schedule, to which reference should be made.

It must be emphasised that the certification of work or an inspection made mandatory by an Airworthiness Directive, or its equivalent, is permitted only by appropriately licensed aircraft engineers or authorised persons in approved companies.

It must be understood that the above privileges apply only in the U.K., to U.K. registered aircraft and to U.K. licensed pilots. It must also be emphasised that these regulations are subject to change from time to time and reference should always be made to the latest issue of the documents mentioned in the text.

Index